FORESIGHT

How do attempts to foresee the future actually change it? For thousands of years, humans have called upon foresight to shape their own actions in order to adapt and survive; as Charles Darwin revealed in his theory of natural selection, the capacity to do just that is key to the origin of species. The uses of foresight, however, can also be applied to help us further our understanding across a variety of realms in everything from warfare, journalism and music, to ancient civilizations, space weather and science. In a thought-provoking new addition to the Darwin College Lecture Series, eight distinguished authors each present an essay from their area of expertise devoted to the theme of 'foresight'. This provocative read reveals foresight as a process that can be identified across all areas of human endeavour; an art which can not only predict the future, but make it anything but inevitable.

Contributors
Geoffrey Lloyd, Bridget Kendall, Robert J. Sawyer, Hasok Chang, Nicholas Cook, Jim Wild, Terrie E. Moffitt and Francesca Rochberg.

LAWRENCE W. SHERMAN is Director of the Institute of Criminology at the University of Cambridge, where he has served as Wolfson Professor of Criminology since 2007. He is also Director of the Jerry Lee Centre for Experimental Criminology and Chair of the Cambridge Police Executive Programme.

DAVID FELLER holds a PhD in the History and Philosophy of Science from the University of Cambridge, as well as a law degree from Georgetown University Law Center, Washington D.C. His work at the universities of Cambridge and of Manchester, particularly regarding use of dogs as models in evolutionary and genetic science, has created new discussion in both Darwinian Studies and concepts of cultural influences in biology. Most recently he has lectured in Law and Science at Cambridge, discussing how law and science combine to guide notions of the future. He is currently the senior contracts manager for the Physical Sciences for the University of Cambridge.

THE DARWIN COLLEGE LECTURES

These essays are developed from the 2013 Darwin College Lecture Series. Now in their twenty-ninth year, these popular Cambridge talks take a single theme each year. Internationally distinguished scholars, skilled as popularisers, address the theme from the point of view of eight different arts and sciences disciplines.

Subjects covered in the series include

27 FORESIGHT
eds. Lawrence Sherman and David Allan Feller
pb 9781107512368

26 LIFE
eds. William Brown and Andrew Fabian
pb 9781107612556

25 BEAUTY
eds. Lauren Arrington, Zoe Leinhardt, Philip Dawid
9781107693432

24 RISK
eds. Layla Skinns, Michael Scott, Tony Cox
pb 9780521171977

23 DARWIN
eds. William Brown and Andrew Fabian
pb 9780521131957

22 SERENDIPITY
eds. Mark de Rond and Iain Morley
pb 9780521181815

21 IDENTITY
eds. Giselle Walker and Elizabeth Leedham-Green
pb 9780521897266

20 SURVIVAL
ed. Emily Shuckburgh
pb 9780521718206

19 CONFLICT
eds. Martin Jones and Andrew Fabian
hb 9780521839600

18 EVIDENCE
eds. Andrew Bell, John Swenson-Wright and Karin Tybjerg
pb 9780521710190

Foresight

Edited by *Lawrence W. Sherman* and *David Allan Feller*

Darwin College, Cambridge

CAMBRIDGE
UNIVERSITY PRESS

CAMBRIDGE
UNIVERSITY PRESS

University Printing House, Cambridge CB2 8BS, United Kingdom

Cambridge University Press is part of the University of Cambridge.

It furthers the University's mission by disseminating knowledge in the pursuit of education, learning and research at the highest international levels of excellence.

www.cambridge.org
Information on this title: www.cambridge.org/9781107512368

© Darwin College, 2017

First published 2016

A catalogue record for this publication is available from the British Library

Library of Congress Cataloging-in-Publication data
Sherman, Lawrence W., editor. | Feller, David Allan, editor.
Foresight / edited by Lawrence Sherman and David Allan Feller.
Cambridge, United Kingdom : Cambridge University Press, 2016. |
Series: Darwin college lectures
LCCN 2016023668 | ISBN 9781107512368 (paperback)
LCSH: Forecasting. | Prediction (Psychology) | BISAC: SCIENCE / General.
LCC CB158 .F67 2016 | DDC 303.49–dc23
LC record available at https://lccn.loc.gov/2016023668

ISBN 978-1-107-51236-8 Paperback

Contents

Figures

Notes on Contributors

Geoffrey Lloyd received the Dan David Prize in 2013 for his contribution to our understanding of the modern legacy of the ancient world; the latest in an august career of academic achievements. Throughout his university career he has been based chiefly at Cambridge, holding various university and college posts, first at King's College and then at Darwin College. From 1983 onwards he held a personal chair in ancient philosophy and science, and from 1989 to his retirement in 2000 he was Master of Darwin College. He is a fellow of the British Academy, and among his many other posts and awards, he has received the Sarton Medal (1987); foreign membership the American Academy of Arts and Sciences (1995); and the Kenyon Medal for Classical Scholarship from the British Academy in (2007). He was knighted for "services to the history of thought" in 1997, having published over twenty books and over one hundred and fifty articles.

Bridget Kendall became Master of Peterhouse, Cambridge in 2016. She joined the BBC in 1983 as a radio production trainee for BBC World Service. She rose to the post of the BBC's Moscow correspondent from 1989 to 1995, witnessing the power struggles in the Soviet Communist Party as Mikhail Gorbachev tried to introduce reform, reporting on the break-up of the Soviet Union and the internal conflicts in Chechnya, Georgia, and Tajikistan. She sent reports of the coup in August 1991 and covered Boris Yeltsin's rise to power. She moved to the BBC's Washington, D.C. desk from 1994, and became the Corporation's diplomatic correspondent in 1998. Kendall speaks fluent Russian and has interviewed world leaders including Vladimir Putin live from the Kremlin as part of an internet webcast in March 2001. She is the host of the talk show *The Forum* on BBC World Service radio.

Robert J. Sawyer is one of only eight writers in history to win all three of the world's top awards for science fiction novel of the year: the World Science Fiction Society's Hugo Award, won in 2003 for his novel *Hominids*; the

ix

Science Fiction and Fantasy Writers of America's Nebula Award, for *The Terminal Experiment* (1996); and the John W. Campbell Award in 2006 for *Mindscan.* His influence in the genre extends well beyond his award-winning example, though, as his work with young writers and public engagement has made him one of Canada's most influential publishing industry figures. Over his twenty-two novels, he has looked at science and its meaning in culture across a vast array of disciplines and applications, and his contributions culminated in October 2013 in the Canadian Science Fiction and Fantasy Association's Lifetime Achievement Award, only the fourth time in thirty years the award has been given to an author.

Hasok Chang is the Hans Rausing Professor of History and Philosophy of Science at Cambridge. His research interests include the history and philosophy of chemistry and physics from the eighteenth century onward; philosophy of scientific practice; other topics in the including measurement, realism, evidence, pluralism, and pragmatism. He was President of the British Society for the History of Science (2012–14) and is a founding member of the Committee for Integrated History and Philosophy of Science. His intriguing publications include *Is Water H_2O? Evidence, Pluralism and Realism* (2012) and *Inventing Temperature: Measurement and Scientific Progress* (2004).

Nicholas Cook took up the 1684 Professorship in 2009. He was formerly Professorial Research Fellow at Royal Holloway, University of London, where he directed the AHRC Research Centre for the History and Analysis of Recorded Music (CHARM), and before that he taught at the universities of Hong Kong, Sydney, and Southampton, where he also served as Dean of Arts. A musicologist and theorist, he holds separate degrees in music and in history/art history. His articles have appeared in leading British and American journals, and cover topics from aesthetics and analysis to psychology and popular culture. His books include *A Guide to Musical Analysis* (1987) and *Music, Imagination, and Culture* (1990). Cook's current work is turning toward social and intercultural perspectives on music, and in 2014 he took up a British Academy Wolfson Research Professorship to work on a three-year project entitled "Musical Encounters: Studies in Relational Musicology." He is a Fellow of the British Academy and of the Academy of Europe.

Jim Wild is a physicist studying the space environment and the links between the Sun, the Earth, and other planets. He has a doctorate in solar-terrestrial physics at the University of Leicester and is now Professor of

Space Physics at Lancaster University's Department of Physics. His research investigates the physics behind the aurora borealis (sometimes known as the northern lights), the impact of space weather on human technology, and the interaction between the Martian atmosphere and the interplanetary environment. In 2010, he was awarded a Science in Society fellowship by the UK Science and Technology Facilities Council. Wild is a fellow of the Royal Astronomical Society and a member of the European Geosciences Union and American Geophysical Union.

Terrie E. Moffitt is the Knut Schmidt Nielsen Professor at Duke University's Department of Psychology & Neuroscience and Department of Psychiatry & Behavioral Sciences. Her main course of research looks at the interplay between nature and nurture in the origins of problem behaviors with particular interest is in antisocial and criminal behaviors. She is Associate Director of the Dunedin Multidisciplinary Health and Development Study in New Zealand, and she also directs the Environmental-Risk Longitudinal Twin Study ("E-risk"), which follows a 1994 birth cohort of 1116 British families with twins. For her research, she has received the American Psychological Association's Early Career Contribution Award (1993), Distinguished Career Award in Clinical Child Psychology (2006), and the Stockholm Prize in Criminology (2007). She is a fellow of the Academy of Medical Sciences (1999), the American Society of Criminology (2003), the British Academy (2004), and Academia Europaea (2005).

Francesca Rochberg is the Catherine and William L. Magistretti Distinguished Professor of Near Eastern Studies at the University of California at Berkeley. Her early academic career included honors as a fellow with the John Simon Guggenheim Memorial Foundation and the John D. and Catherine T. MacArthur Foundation. Her current research focuses on Assyriology, with an emphasis on Akkadian scholastic texts of the second and first millennia BCE, Babylonian astronomy and astrology, and the impact of the philosophy of science on the historiography of ancient science. In 2010, she was posted at the Ludwig-Maximilian Universität, München as a research fellow, and is a member of the American Philosophical Society. In 2007 she became a member of the Princeton, Institute for Advanced Study, School of Historical Studies, and has received the John Frederick Lewis Award for *Babylonian Horoscopes* (American Philosophical Society, 1998).

Introduction: The Foresight Dialectic

LAWRENCE W. SHERMAN AND DAVID ALLAN FELLER

The idea of foresight is as old as recorded societies, as controversial as climate change, as interesting as a murder mystery, and as useful as an umbrella. It is a siren tempting journalists to dash themselves on the rocks of prediction, a competitive sport in which winners may claim high fees for their success—or even failure. Yet the greatest value of foresight is to *change* our futures, not just to predict them. Foresight embraces more than disengaged claims about what will happen; it can also engage lines of action in which foresight itself causes what happens next—either to confirm its prediction or avoid it. As Francesca Rochberg states in Chapter 8, "foresight is not simply looking forward, but anticipating future change and acting on that vision."

Dialectic

The dialectic between prediction and causation runs throughout this volume, across disciplines as diverse as journalism, music, and ancient history. The consistency of that theme was neither foreseen nor required by the organizers of the 2013 lectures documented in this volume, the twenty-seventh in the renowned Darwin College series presented in Cambridge every year since 1986. The convergence of eight independent thinkers on this dialectic was emergent, unplanned, and serendipitous. As the result of what turned out to be an intellectual Rorschach test, presenting the idea of foresight repeatedly led our authors to interpret it as a blend of prophecy and its fulfilment.

While no author mentions the idea of "pre-crime" explored in the 1956 Philip K. Dick story *The Minority Report*, in which foreseeing a future crime in the brain of the would-be murderer allows police to prevent the

1

murder, Robert Sawyer, in Chapter 3, comes close. He argues that for foresight in fiction to "extrapolate to what plausibly might happen," the first requirement is the author's "conviction that human nature does change, that our psyches and our societies are malleable." And so, he implies, we have and will act to change them: "It is this ability to change that may explain why we're here and all other forms of humanity have died out." Charles Darwin would agree.

Even when it is wrong, foresight can shape the future. Bridget Kendall, in Chapter 2, reviews the consequences of the 1987 television announcement that there would be no hurricane the next day, when there was. Without counting the exact cost of this error, it is possible to accept that more preventive measures from an accurate forecast could have reduced the losses which the hurricane caused. The long-run way in which that error may change the future is more beneficial: in 2014, the UK Meteorological Office announced major investment in supercomputers and satellite data collection that can increase the micro-regional precision of forecasts of floods, high winds, and other weather disasters early enough to reduce their damage. The journalism reporting this new capacity comprised a Greek Chorus chanting reminders of the 1987 error.

Thesis and Antithesis

Our authors are anything but one-sided in the link between foresight and outcomes. They do not claim merely that accurate foresights cause action that changes the future, or that inaccurate foresight causes inaction that leads to disaster; that two-sided coin is merely the thesis in the dialectic. The antithesis in this dialectic is that *even inaccurate or fabricated foresights can create a self-fulfilling prophecy* in which the foresight becomes accurate only because it was proclaimed with great conviction.

Francesca Rochberg, in Chapter 8, observes this phenomenon in ancient Mesopotamia, in the rise to power of Esarhaddon (681–669 BC). "According to the Assyrian imperial master narrative," she writes, "Esarhaddon told of his own appointment as crown-prince by divine selection." He then cited specific omens as evidence of the decision of not one, but many, Gods that he should become the next ruler. Using these omens as a weapon to defeat his brothers and others vying for the throne, Esarhaddon's "announcement of the omens was an essential ingredient of the master narrative, the key to securing a legitimate claim" to rule.

It is easy to scoff at anyone's claims of divine selection to rule, as the Stuart Kings of England discovered after James I offered an in-your-face insistence on the sacred nature of his family's Royal powers. But just as foresight of divine intentions can create self-fulfilling prophecy, so can scoffing at foresight itself. As Terrie Moffitt suggests in Chapter 7, a strong lack of interest in foresight can spell disaster for individuals with that predilection stretching over their entire life course.

Foresight → Better Outcomes

Analyzing the life histories of 1037 people born in Dunedin, New Zealand in 1972–73, Moffitt shows that low self-control is strongly associated with a wide range of poor outcomes that can be prevented by foresight, including financial foresight. By their mid-thirties, people who had low self-control measured from ages three to eleven were clearly less "financially planful." Compared to their peers with better self-control, they were "less likely to save and they had acquired fewer financial building blocks for the future (such as home ownership, investment funds, or retirement plans)," a finding that was clear even when IQ and parental social class were held constant. These same impulsive people were also seen in a statistical gradient, in which the less self-control they had, the less foresight they had shown in many aspects of life: health (including smoking), wealth, parenting skill, staying free of criminal prosecution, getting an education.

The point of Moffitt's analysis fits right into the foresight dialectic. While she demonstrates a strong link from low foresight to poor outcomes, she repeatedly claims that low foresight is malleable. If societies have foresight about members with low foresight, she argues, we should be able to intervene at many stages of life to help them take foresight more seriously. This takes us from the anti-thesis back to the thesis, as Robert Sawyer argued, that we prosper as a species because we can change ourselves to adapt better than we have in the past.

Foresight → Worse Outcomes

Yet this thesis must include the argument that societies may use foresight to change themselves for the worse, as Nicholas Cook reminds us in his analysis of foresight in music in Chapter 5. Building on Jacques Attali's claim that "every major social rupture has been preceded by an essential mutation in the codes of music," Cook pursues the link between the musical foresight and the success of the change it predicts. He cites analyses of the Balkan wars of the 1990s as evidence that music helped to promote the conflicts, including Donna Buchanan's claim that, performances and recordings by Bulgarian musicians during the 1980s "functioned not only as prominent harbingers, but as agents of political transition." Did foresight predict the struggles, cause them, or both?

Cook's lecture carries the analysis further, to consumers of music becoming its producers in a kind of crowd-source production by consumers, or "prosumption." Foresight in this sense becomes an iterative process of innovation provoking (causing) more innovation that repeats and intensifies the production of a new kind of music. Thus foresight in the prosumption of music is changing music itself, while the new music is changing its wider social context. Whether that context is better or worse than its predecessor is beyond the boundaries of the thesis, which allows a wide range of other factors to affect the good of the outcome.

No example of foresight causing events for the worse is more vivid than Bridget Kendall's discussion, in Chapter 2, of the 2012 death of reporter Marie Colvin in Syria. Noting the dilemma of journalists offering foresight in the middle of ongoing events, Kendall places the circumstances of a journalist's death in the wider context of all journalism in covering events in which the main message is that people are dying, and innocent civilians are being slaughtered: "If you are crouched in a bombed out house in a suburb of Homs during the Syrian conflict, like the late *Sunday Times* correspondent Marie Colvin and her colleagues in 2012, should you steer clear of giving live interviews by satellite from your location? It is thought that it may have been the Syrian military picking up her signal and launching an attack on the house where she was staying which led to her death."

Hasok Chang, in Chapter 4, also tackles the problem of foresight being right in its prediction, but uncertain in the effects of making the prediction. This problem is not essentially dialectical, since it starts by simply drawing boundaries between predicting two different phenomena (the subject of interest—like the weather—and the effects of the prediction—like closing schools, or not). But it has major implications for the role of science in the foresight dialectic, which may be to claim exemption from a more general view of foresight. When it comes to foresight in science, Chang argues, success in prediction has never been certain to lead to further success. The history of science shows a wide range of sequelae of predictive success, with foresight about science itself having a miserable track record.

Error and Humility

Chang's argument about foresight in science echoes Kendall's central concern about foresight in journalism, which is that "foresights" are so very often wrong. Both authors counsel humility in offering foresight, if only because there are so many unimagined alternative scenarios that may still come to pass. Chang quotes Joseph Priestley saying in the eighteenth century that accurate predictions may expand ignorance even while expanding knowledge: "every discovery brings to our view many things of which we had no intimation before." Chang also quotes a wonderful image that Priestley offered for this claim: "The greater is the circle of light, the greater is the boundary of the darkness by which it is confined."

Yet is this true for every area of science? In Chapter 6, Jim Wild implies that progress has reduced ignorance while expanding knowledge of "space weather," defined as encompassing "conditions and processes occurring in space, including on the sun, in the magnetosphere, ionosphere and thermosphere, which have the potential to affect the near-Earth environment." Wild reports that the planetary systems affecting space weather are quite stable, with little change to foresee in the short or long term. Foresight's problem is to understand how new human technologies may ·be disrupted by foreseeable bouts of space weather disruptions.

The 1859 geomagnetic storm, known as the "Carrington event," provides only a glimpse of what could happen today. The telegraph was the only technology affected by that event. But a similar storm today could knock out all satellite-generated technologies, from air traffic control to satnav guidance in automobiles to GPS tracking of sex offenders wearing electronic shackles on their ankles. At worst, it could knock out the electrical grids for entire nations, as in the experience of all Bangladesh in late 2014. Hospitals, heating systems, and water and food supplies could all be disrupted. Wild reports estimates that full recovery from another Carrington event could take four to ten years.

Wild's account of these hazards calls for better forecasting of such events. But true foresight about them would be much broader, at least in the foresight dialectic implied by the other authors. The thesis of that dialectic would be that damage from space storms can be minimized by humans refining technologies that humans designed. A big vision, such as giant surge protectors for every use of electricity, might or might not be useful, but something creating that effect would be very useful. In foreseeing these hazards for such a purpose, they could be prevented. Foresight can falsify the predictions by making them not come true—but only if humans adapt and take corrective action.

Mistakes → Improvements

Yet as the climate change debate demonstrates, attempts to promote such foresight may reveal a modified antithesis—that *even inaccurate or fabricated foresights can create a self-fulfilling prophecy.* Instead of the foresight becoming accurate only because it was proclaimed with great conviction, an equally convinced *denial* of the foresight can also make the foreseeable horrors come true. Financial interests tied to short-term denial of long-range foresights, or foresights of uncertain dates, can persuade democracies there is nothing to worry about. The collapse of human civilizations from Greenland to Easter Island may offer mute tribute to the success of such foresight deniers.

This volume begins with several ancient civilizations, including China, the Graeco-Roman world, and Mesopotamia. In Chapter 1, Geoffrey Lloyd considers four questions about foresight in those societies, the last of which is central to the foresight dialectic: "How is foresight related first to divination in particular, and prediction in general, and then to wisdom and prudence?" The word "prudence" is central, as Terrie Moffitt agrees in Chapter 7, since it denotes the fusion of prediction and prevention, decisions made to avoid harm and promote survival – exactly what *individuals* with the weakest self-control fail to do. But how does foresight shape prudence in entire *societies?*

Lloyd's answer is that prudence in ancient civilizations might require ample scepticism about the standard methods of predicting events. He reports that a first-century BCE Chinese philosopher named Wang Chong challenged the methods of his day in concluding that "The widespread opinion that the dead turtle shell and the dried milfoil stalks can … obtain replies to questions that are put is erroneous." Lloyd also cites Sunzi's treatise on the art of war, emphasizing "the need for foreknowledge, but says that that is not to be got from ghosts and spirits, nor by comparing past events, nor from plotting the positions of the heavenly bodies: no: it can only be got from knowing the enemy's situation." This means spies—a particularly non-divine source of foresight.

Lloyd cites sceptics in other ancient civilizations, showing that scepticism about irrational methods was itself rational. But what if the methods had been firmly grounded in science? Our most serious modern challenge to prudence from foresight is public scepticism about strong evidence. Yet that challenge may come from Lloyd's distinction between precise predictions and more general scenarios plotting likely outcomes of alternative courses of action. The predictions were largely based on non-empirical divinations, but the scenarios were derived from empiricism of history. Lessons learned from histories of past mistakes informed the wisdom of China, Greece and Rome. The mistakes were tied to stable features of human societies at that time, including governance, warfare, food supplies, and crises.

Today's crises may be fundamentally different, even if human nature is not. We face many problems for which the past provides little guidance. From magnetic space-storms to climate change, human effort is changing the nature of the problems which humans face. The number of humans on the planet is unprecedented, which makes even the oldest of problems unprecedented in their scale. Our sceptics may have evolved to look to the past to find wisdom for prudent actions in foresight. Much wisdom, undoubtedly, can still be drawn from the past. But the past can hardly be a complete guide. What the foresight dialectic tells is that adaptation is needed to prevent a harmful future. That, in turn, requires that we capitalize on Robert Sawyer's conviction that human nature, and human societies, are malleable.

Hope or Humility?

Here is where our authors part company, since Sawyer's conclusion challenges the counsel for humility offered by Bridget Kendall and Hasok Chang. Chang concludes that "true foresight consists in recognising ... proper limits." Nicholas Cook, Terrie Moffitt, Francesca Rochberg, and Robert Sawyer hold out the hope for an end to the curse of Cassandra. Yet all authors agree on the power of foresights—right, wrong, or conditional—to influence decisions—right, wrong or disastrous. Humility is wise, but so is empiricism. The empirical evidence assembled in these eight lectures offers powerful evidence for the contingent, probabilistic nature of both predictions and prudence, changing conditions and human adaptations.

These characteristics of foresight go well beyond the making of predictions. They mean that each statement about what the future will be can provoke a dialectical discussion over what the future should be, and ultimately becomes. Whether that idea brings the reader hope or despair is the ultimate Rorschach test of this volume.

The Darwin College Lectures

That such agreement on these choices can be achieved across a wide range of topics and disciplines is yet another tribute to the Darwin College Lecture series. The very idea for the Darwin Lectures was a case of foresight becoming a self-fulfilling prophecy. The foresight was that in a world of increasing specialization and differentiation of disciplines, there would be continuing intellectual interest in a multi-disciplinary view of big ideas and subjects. Since that claim was stated, and especially since videos of the lectures have been posted online for some one million views, Darwin College has helped to sustain the broad intellectual approach it foresaw.

C.P. Snow observed that science was rarely discussed at Cambridge college dinners, since the divide between science and humanities is so great that "there seems then to be no place where the [two] cultures meet." Yet for three decades, Darwin College has created just such a meeting in Cambridge on Friday nights of every January and February. It has even added the third culture, of social science, to this gathering of minds and perspectives.

Every Darwin College Lecture series must touch all three realms of thinking, including physical science as well as its less-exact siblings. This mandate is not only challenging to the convenors of each series, but also to the readers of each volume. It is tempting to think that readers may sit down and plow their way through such volumes from cover to cover. A more modest ambition is that any reader might enrich their lateral thinking by reading even one lecture on a topic of interest by someone from an entirely different field. This is, after all, what Steven Johnson[1] argues to be the best source of new ideas, just as Gutenberg devised the printing press after viewing a wine press in action. After almost three decades, the Darwin College Lectures have become established as a reliable venue for lateral thinking from cross-disciplinary insights—and sometimes even foresights.

---

[1] *Where Good Ideas Come From* (N.Y.: Riverhead Books, Penguin Press, 2010).

1 Foresight in Ancient Civilisations

GEOFFREY LLOYD

> Good health to the king, my lord! May the Gods Nabû and Marduk bless
> the king, my lord. Concerning the watch of the moon about which the king,
> my lord, wrote to me, (the eclipse) will pass by, it will not occur.
> Concerning the watch of the sun about which the king, my lord, wrote
> to me, does the king, my lord, not know that it is being closely observed?
> The day of tomorrow is the only day left: once the watch is over, this
> eclipse, too, will have passed by, it will not occur.
>
> (Parpola 1993)

> The 13th day, the night of the 14th day, is the day of the watch to be held
> and there will be no eclipse. I guarantee it seven times: an eclipse will not
> take place. I am writing a definitive word to the king.
>
> (Hunger 1992)

> He who wrote to the king, my lord, 'The planet Venus is visible, it is visible
> in the month Adjar (XII)' is a vile man, an ignoramus, a cheat! And he who
> wrote to the king, my lord, 'Venus is . . . rising in the constellation Aries'
> does not speak the truth either. Venus is not yet visible!. . . Why does
> someone tell lies and boast about it? If he does not know, he should keep
> his mouth shut.
>
> (Parpola 1993)

These three tablets (exhibits 1–3 above) come from letters to the kings of
Assyria, collected in the marvellous State Archives of Assyria series. They
were written sometime around the second to fourth decades of the
seventh century BCE by scribes (they were known as *ṭupšarru*) and sometimes we know their names. The first comes from a scribe named Balasî,
the third from Nabû-ahhe-eriba. They are not the oldest material I shall
be setting before you, but they allow me straight away to identify the main
topics on my agenda.

1   What items are the subjects of prediction?
2   How were predictions made? On what basis did people think they could foresee the future?
3   Why were they interested in doing so?
4   How is foresight related first to divination in particular, and prediction in general, and then to wisdom and prudence?

I should explain that I shall be concentrating mainly on Mesopotamia, China and the Greco-Roman world. I am not concerned here with the tricky issues of who might have transmitted what to whom, and when; and, besides, some of the ideas and practices I shall discuss are unique to just one ancient culture. Rather, I am concerned to mine the very rich materials from those three ancient civilisations to see what light they can throw on my four strategic questions.

Subjects of Prediction

We might assume that the desire to see into the future is as old as the human race. Yet we should be careful. Some societies are far more focussed on the immediate present, on the here and now, and don't appear to plan long term. Conversely, others (India in particular) have a notion of immense cycles of millions of years, the *kalpas*, sometimes combining that with a belief in 'eternal return'. That was sometimes the belief: not just that similar sequences of events recur, but that identical individual events do. Sometime in some remote future in some world someone will be lecturing on foresight in ancient civilisations, and waving his hand just like me. I adapt that image from what the fourth-century BCE Greek philosopher Eudemus tells us about the notion of cyclical time that was held by certain Pythagoreans. According to Simplicius' *Commentary on Aristotle's Physics* 732.30ff., he said, 'If one were to believe the Pythagoreans, that the same individual events recur, and I shall be talking to you, holding my stick, as you sit there, and everything else will be as it is now, then it is reasonable to say that time repeats itself.' But the majority of people in most societies have been much concerned with their immediate and short-term prospects. How many of you consult your horoscopes? Of course you don't believe what you read. But why do you do it? Out of curiosity? But on what basis?

Will I pass my exams? That is an appropriate one to begin with at Cambridge, though it is not just a problem here but a major preoccupation in modern-day Japan (for instance) to judge from the thousands of prayers one finds on the subject outside of shrines – but it was also a concern in the civilisation that invented the examination system, imperial China. Then what will my children become when they grow up? Tinker, tailor, soldier, sailor: rich man, poor man, beggar man, thief. That's a bit dated, but we used to do that in my youth with cherry stones, when we were lucky enough to have cherries. Is this boy or girl a suitable person for my daughter or son to marry? Will the baby be a boy or a girl? Nowadays there are tests that purport to predict that, though two of my grandsons were said to be granddaughters before they were born. More startling still, when my own second son was born, the very nice, well-educated, properly trained Cambridge midwife, felt my wife's womb just before delivery and said 'It's going to be a boy.' I asked her how she knew. 'Well he's on the right.'

That's an idea that can be traced to the fifth-century BCE Greek philosopher Parmenides. He is famous for having produced what he represented as a knock-down argument to show that change and coming-to-be are impossible, so what was he doing talking about childbirth, predicting the sex of the unborn baby? Nevertheless, he is reported as having said, 'On the right boys, on the left girls.'

My next exhibit comes from the oracle bones dating from 1200 or so BCE from Anyang in China: 'Crack-making on day 39, Que determined: "Lady Hao will give birth and it will be advantageous" ... Crack-making on day 39, Que determined: "Lady Hao will give birth but it will not perhaps be advantageous."' Usually, but not it so happens in this case, an auspicious sign meant a male, an inauspicious one a female, child.

Crack-making was a matter of causing cracks by firing turtle shells (or bones) and getting an answer to a predetermined question called 'the charge', by interpreting the pattern the cracks made. Figure 1.1 shows a turtle shell used for twelve pairs of charges. Que was the diviner in charge of the crack-making in the last slide, but it is the king who makes the prognostication. And note the question is put in both a positive and a negative form ('it will be advantageous', 'it will not perhaps be

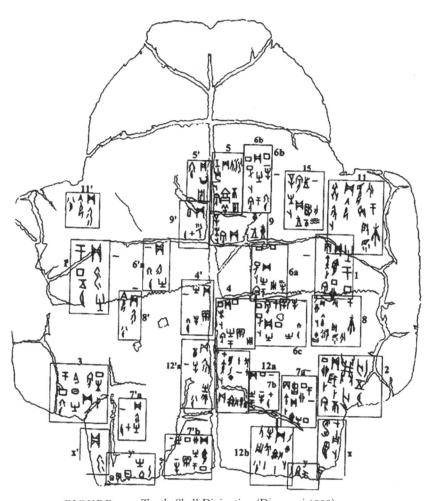

FIGURE 1.1 Turtle Shell Divination (Djamouri 1999)

advantageous'). But for now I have introduced this exhibit just to show that determining the sex of the unborn child has quite a track record in terms of recurrent human concerns.

We now have the beginning of the answer to my first question (what items are the subjects of prediction). The Chinese and the Mesopotamian data above show they were particularly preoccupied with whether to go to war or not – as also were the Greeks. Remember King Croesus, the king of Lydia, who took the trouble to consult the Delphic oracle about whether to invade Persia. Herodotus (I 46–53) reports that he first tested all the main oracles in Greece, Delphi, Dodona, the shrines of Amphiaraus and Trophonius, and Branchidae, as well as the Libyan oracle at Ammon, to determine how good each oracle was. He asked them to say what he was doing on a particular day, when there was no way by which they could have known: Delphi passes the test with flying colours; he was cooking a tortoise and lamb stew in a bronze caldron. Actually our source, Herodotus, says that as soon as his emissaries entered the hall at Delphi, the Pythian priestess broke into hexameter verse, claiming that she could smell what was cooking miles away in Sardis. This startling case of what we might call tele-olfaction convinced Croesus that this was the oracle to go for and that he could trust. When he sent more emissaries with loads of gifts to Delphi and put forward the question: 'Should Croesus send an army against the Persians and should he collect allies to do so?' The answer was: 'If he should send an army against the Persians, he would destroy a great empire.' The trouble was that the empire Croesus destroyed was his own. (Remember that the fourth question on my agenda which I shall come to is the relation between foresight, prediction, wisdom and prudence.)

It does not require great skills of divination on my part to arrive at the conclusion that the items about which foreknowledge is sought tally with everything that was of concern to the individuals or groups involved – so predictions are useful evidence for those concerns and the values of the groups in question. That is one of the reasons why they are so well worth studying. Predictions are revealing evidence as to what exactly was on people's minds. Kings have their particular interests of course: war, illnesses, journeys, births, intrigues at court, sexual partners, hunting; but groups that actually subsist on hunting also have to decide where and when to set out. (I shall come back to that.) And of course it is not just important events that occupy us, but plenty of mundane ones. Betting on the outcome of chariot races was as much of an obsession in Byzantium in the fourth century CE as doing the football pools or the lottery are today.

Methods of Prediction

But then my second, more challenging question is: How were predictions made? On what basis did people think they could make them? The answer is importantly complex, and I shall approach the problem indirectly.

There are hundreds, maybe even thousands, of different techniques of looking into the future that we know about from ancient and modern societies. Watching the movement of birds is a favourite in Mesopotamia, China and Greece. Opening up animals and looking at their viscera, especially the liver and the gall bladder – extispicy – is another common one. Figure 1.2 shows one of a number of extant models of livers from Babylonia (this one now in the Science Museum in London). One idea is that the models were used to teach apprentice diviners how to read 'the signs'. How did they do that? They learned it out of books or from their teachers, neither of whom you could really call into question – not if you wanted to become an accredited diviner yourself.

Dreams were another very frequent source of ideas about what was to happen, though as Homer already pointed out some were veridical, some not: there was a Gate of Horn for dreams that will come to pass, but also a Gate of Ivory for those that will not. But in literate societies people also used books. In pagan antiquity there was the famous Sortes Virgilianae:

FIGURE 1.2   Babylonian Model of a Liver Used in Divination

you opened the text of Virgil at a particular line and what you found was supposed to be significant. Christians were to do that a lot with the Bible. To ancient techniques we can add many methods attested from present-day societies: the famous poison oracle among the Azande (where, like the Chinese oracle bones, the question was put twice, once in a positive and once in a negative form: if the results were inconsistent, you discarded them) and the intricate interpretation of the tracks that a fox made on a marked area of sand during the night (as reported among the Dogon). The Greeks and then the Romans (Cicero for instance) tried to establish a broad division between what they called artificial or external divination and natural or internal divination. The artificial variety depended on inspiration: prophets needed that, and you needed that to interpret dreams. But natural divination depended on art and knowledge. The interpretation of bird signs and of celestial phenomena fell into that category: you had to know what your learned predecessors had discovered from their experience or from what they said they had.

Now let me quote another excerpt from the Letters to Assyrian kings: 'If the Pleiades enter the moon [that is to say they are occulted by the moon] and they come out to the north, Akkad will become happy: the king of Akkad will become strong and have no rival' (Hunger 1992, 250). Well, happiness is a pretty indeterminate result, so the diviner could hope for the best that it would turn out all right. The next tablet is much more specific and dangerous: 'If Jupiter becomes visible in the path of the Anu stars, the crown prince will rebel against his father and seize the throne' (Parpola 1993, 299). The whole of astrology gets built up on presumed correlations between signs and outcomes. Horoscopes go into considerable detail on the topic. Well, some do. There is a distinction between very basic predictions, where just the positions of the sun and moon are taken into account, and what Alex Jones, the expert on Greek horoscopes, has called the deluxe models (much more expensive), which go into great detail on the relative positions of all the heavenly bodies. There is still a good deal of room for manoeuvre to allow the astrologer to adapt his account to the person he is dealing with. One of my ex-graduate students, Tamsyn Barton, is responsible for a brilliant handbook on the subject, in which she exploited just that feature. First she did a horoscope for the then Professor of Ancient History (he was not amused). Then she gave a result for Prince Charles, which long before the tragic death of Diana, spoke of Charles' problems with his love life (see Figure 1.3). Tamsyn would have been a great success in the ancient world, but she would have been lucky to have got away with such a prediction with her life.

But by far and away the most elaborate of all methods of divination (I am sure you have been waiting for it) is in the Chinese *Book of Changes*. Turtle shells (such as we saw used in Figure 1.1) were pretty costly items. The *Book of Changes* had the great advantage that the diviner used sticks of yarrow or milfoil – much cheaper. Actually, since the lines could be associated with odd and even numbers, you could do it – as many still do – with coins (heads and tails). In the original version, the diviner used a bundle of sticks, and by a programme of sorting and discarding some,

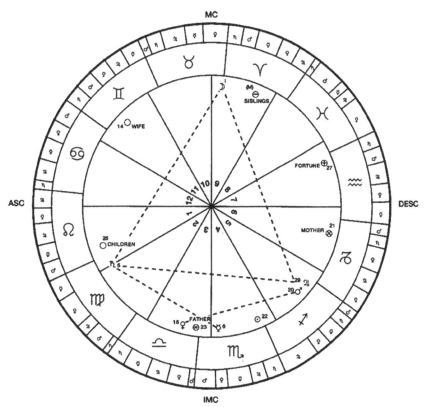

FIGURE 1.3 Tamsyn Barton's Horoscope for Prince Charles

Hexagram "Caldron".

The Judgement:
Supreme good fortune. Success.
The Image:
Fire over wood; the image of the caldron....
The Lines:
Six at the beginning means:
A Caldron with legs upturned.
Further removal of stagnating stuff.
No blame.
Nine in the second place means:
There is food in the caldron.
My comrades are envious
But they cannot harm me....
Nine in the fourth place means:
The legs of the caldron are broken.
The prince's meal is spilled
And his person soiled.
Misfortune.

FIGURE 1.4 Hexagram Caldron

ended up with an array of six broken or unbroken lines, a hexagram. Figure 1.4 gives the hexagram that came to be called *Ding*, the caldron. The text first gave the 'judgement', in this case 'Supreme good fortune. Success', then the image, then remarks on each of the lines, and the whole book was supplemented by extensive commentaries, the so-called Ten Wings. The diviner read the hexagram from the bottom up, but the next complication is that some of the lines are unstable and switch into their opposites, so that the original milfoil array gives not just one, but several possible situations. And note that in this particular case, although the overall judgement is auspicious, one of the lines ('nine in the fourth place') is associated with misfortune. You can never be too careful.

By now your heads may be reeling, and you may be saying to your-selves that all this material that I have been putting before you is just dismal testimony to human self-delusion, to our capacity to kid ourselves. It is no doubt natural, you will agree, that we would like to know the future. But in the vast majority of cases the techniques used have been hopeless, and anyone who believed in their usefulness was up a gum tree.

If that thought occurred to you, you are not the first. My next text comes from the first-century BCE Chinese philosopher Wang Chong. What is so special about yarrow or milfoil that sticks of the stuff can yield predictions? What is so special about the shells of turtles? They are dead and the dead cannot communicate with the living. So he concludes: 'The widespread opinion that the dead turtle shell and the dried milfoil stalks can make contact with heaven and earth and obtain replies to questions that are put is erroneous' (Lun Heng 71, juan 24). Good sceptical stuff. Yet Wang Chong hesitates. He still thinks that it is the case that fortunate people happen to have auspicious omens, and unfortunate ones have inauspicious omens. So if anyone claimed (as many did) that experience proved that there was something to these predictions, Wang Chong can explain that away, though he had no real account to offer of how that correlation happened. But he knows of cases where rulers and generals ignored the omens and got away with it. He cites the case of King Wu who destroyed the kingdom of Zhou. He consulted the turtle and the milfoil, and the signs were adverse and the prognostication was given 'highly inauspicious'. But he just threw the stalks away and trampled on the turtle saying, 'How can dried bones and dead grass know what is inauspicious?' Similarly, in the great treatise on the art of war by Sunzi, we find a text (ch. 13) that emphasises the need for foreknowledge, but says that it is not to be got from ghosts and spirits, nor by comparing past events, nor from plotting the positions of the heavenly bodies: no: it can only be got from knowing the enemy's situation. And how do you get that? You use spies – which leads Sunzi into a discussion of the five kinds of spies to achieve that end.

There are plenty of sceptics around in our other ancient civilisations too. Some Mesopotamian scribes accused others of being cheats and frauds. In Greece we have a Wang Chong-like event in Homer. In book twelve of the *Iliad* (200ff., 211ff., 231ff.) the seer, Polydamas, interprets a bird omen, in this case the fight between an eagle and a snake, where the snake gets away. Polydamas sees this as predicting that the Trojans (who at that point have pushed the Achaians right back to their ships) will be thwarted of final victory. But this is greeted by Hector with: 'one omen is best: to defend one's country'. But look at the complications. First, Hector gets *his* confidence from the promises of victory he says he has had from

Zeus. Second, Polydamas actually turns out to have got it right. The Trojans don't capture the Achaian camp.

But all that scepticism just shows, does it not, that there were some brave souls who agreed that most techniques of divination were hopelessly deluded. But that does not apply to them all. Two points here. First, as some modern anthropologists have pointed out (I am thinking of Bascom, of G.K. Park and the wonderfully named Omar Khayyam Moore) when a particularly tricky decision has to be taken and it would be invidious for an individual to be solely responsible, an impersonal mode of divination may be used as the basis for a consensual verdict as to what to do. In what direction should the hunting party set out? Divinati on can resolve the issue without implicating any individual with blame for failure (or allowing anyone to claim credit for success). Such a procedure introduces a random element in the course of action to pursue, and over the long term the hunting may be more successful when the party does not always follow the same pattern.

But in that case the agents' own beliefs are that they are getting positive signs from some personal or impersonal agency. What we need now is to consider those cases where the prediction is in fact justified by the evidence. That was why I started with a Mesopotamian eclipse prediction. Mesopotamian astrology started with predictions about events on earth on the basis of certain celestial phenomena (where the connections were those that the divinatory handbook the *Enūma Anu Enlil* set out). The scribes were lucky if some of their predictions turned out to be correct, though in predicting whether the baby was to be a boy or a girl you had a 50 per cent chance of success, more or less. But then by the time that most of the Letters I cited were composed and in some cases even before, the scribes were in a position to predict certain celestial phenomena – and to get them right. The scribe in exhibit two was absolutely confident that the eclipse would not take place. And sometimes in the much more difficult case of predicting *that* an eclipse will occur, they get that right too. One tablet reads: 'On the 14th day the moon will make an eclipse. It predicts evil for Elam and the Westland, good for the king, my lord ... Already when Venus became visible, I said to the king, my lord "An eclipse will take place"' (Hunger 1992, 222, 224). Thanks to extensive observation of eclipse cycles carried out over centuries, they had good grounds for their predictions at least so far as lunar eclipses go. Solar eclipses are more difficult, because of their varying visibility from different locations on the surface of the earth, though some claim they could do them too. One scribe writes: 'Concerning the solar eclipse about which the king wrote to me "will it or will it not take place? Send definite word", an eclipse of the sun, like one of the moon, never escapes me' (Parpola 1993, 130).

So here is a kind of case where the diviner could answer the sceptic. This is not just hopelessly fanciful correlation, but based on facts. We tend to hail such achievements as major scientific breakthroughs. But we have to pay attention to the very different perspective that the ancient *ṭupšarru* had on them. They could indeed feel confident over a range of their predictions. But recall I asked *why* they were concerned (question three on my agenda). Of course, if it was a matter of the much-more-difficult subject of medical predicting, of saying whether a patient would recover from illness or not (where some ancient practitioners built up extensive data-bases correlating the signs of the pulse for instance and the outcome of the complaint), in that context there is no great difficulty in identifying the answer to that *why* question. We all want to know what we are suffering from, and whether we shall recover, and even more of course we want to receive treatment that will increase our chances of recovering.

Motives for Prediction

But why would the ancients be bothered about drawing up lists of eclipse cycles? They were not concerned with getting into the history books as the discoverers of celestial periodicities, the discoverers of astronomy we could even say, as opposed to astrology. Not at all: their interest is *still* in what the eclipse *foretold*. Even after they were in a position to predict one, they still believed that an eclipse was an omen – a sign that the king should pay attention to, as indicating a danger to his person or to the realm.

There are two connected parts to the answer to the *why* question where the Babylonian *ṭupšarru* are concerned. If the king was warned of an impending eclipse, then he could and should take evasive action. The ritual known as *namburbû* involved putting someone else (usually a condemned convict) on the throne temporarily, so that whatever evil happened would happen to the substitute king, rather than the real one, who was meanwhile referred to not as king but as 'the farmer'. One occasion on which this happened is recorded in my next tablet. 'To the farmer, our lord, your servants Adad-šumu-uṣur and Marduk-šakim-šumi: Good health to our lord! May Nabû and Marduk bless our lord. Concerning the 15th day about which our lord said "Let the substitute king go to his fate and let me perform my ritual on the 16th day as before", the 16th day is a good day to perform the ritual' (Parpola 1993, 174). You see all sorts of assumptions about causation are in play here. It seems that whoever or whatever was responsible for the evil occurrence was taken in by the substitution: as if it were the throne itself that was some kind of lightning conductor, though it conducted the evil *to* the throne, not away from it: so provided you did not actually sit on the throne yourself, you were safe.

But the second part of the answer to the *why* question concerns the relationship between the *ṭupšarru* and the king himself. When scribes could tell the king what to do, that showed how seriously their predictions were taken. When they got it right, their prestige naturally grew enormously. That is why so much was at stake between one scribe with one prediction and another scribe with a different pediction. He who told the king my lord (we saw in exhibit three) that Venus is visible, is a cheat, an ignoramus: he should not boast but keep his mouth shut. My next tablet was given the modern title: 'the king must give up fasting'.

> To the king, our lord, your servants Balasî and Nabû-ahhe-eriba [our old friends]. Good health to the king our lord! May Nabû [the god after whom the second scribe was named] and Marduk bless the king, our lord.
> The king, our lord, will pardon us. Is one day not enough for the king to mope and to eat nothing? For how long still? This is already the third day, when the king does not eat anything. The king, a beggar! Surely when, in the beginning of the month, the moon appears, he says: 'I will not fast any more. It is the beginning of the month. I want bread to eat and wine to drink!' ... The king can ask for food for even the whole of the year!
> We became worried and were afraid, and that is why we are now writing to the king.
>
> (Parpola 1993, 33)

We see the battles for prestige and influence at work. On the one hand these observers of the heavens and of other natural phenomena as we call them were indeed intent on trying to get it right. But on the other hand, what was at stake was reputation (as indeed it still is). Medical prognostication was much more difficult than celestial. But we see the reputation element at work here too. The fifth-century BCE Hippocratic treatise called *Prognostic* tackles the problem directly in Chapter 1.

> It seems to me to be an excellent thing for a doctor to practise forecasting. For if he discover and declare in advance by the side of his patients [i.e. without his patients telling him] the present, the past and what will happen in the future and supply what the sick have omitted in their accounts, he will be the more believed to understand the problems of the sick, so that people will confidently entrust themselves to the doctor. Furthermore he will carry out the treatment best if he knows beforehand from their present sufferings what will happen later. Now to make every sick person well is impossible. To do so indeed would have been better even than forecasting the future [Good for him, but then he reverts to his usual preoccupation] ... In this way you will rightly be an object of admiration and be a good doctor. For the longer time you plan in advance to meet each contingency the greater your ability to save those who have a chance of recovery, while you will be blameless if you learn and declare beforehand those who will die and those who will get better.

It is pretty optimistic, one might say, that the doctor will be blameless if he has predicted the patient's death: but the fact that that point is made illustrates how, from the doctor's point of view, it is his reputation he is concerned about. He needs to do the best job he can in making his prognosis, in part at least because that will build up the confidence of his clientele.

Foresight, Divination, Prediction, Prudence and Wisdom

So we come to foresight and the fourth, main, question on my agenda, which is the relations between foresight, divination, prediction, prudence and wisdom. At this point we have to get a bit analytic, but if I have carried you so far, I hope I can take you a little further. Prediction I take to be the generic term for being able to make statements about what will happen, useful in all sorts of contexts and evidently a recurrent human preoccupation. Divination is one mode of prediction, used especially where the connection between the evidence used for the prediction and the result that is actually predicted may be opaque, a matter of convention, based simply on what tradition has laid down by way of correlations between signs and outcomes. Another Hippocratic author is relevant here, the author of the treatise *On Regimen in Acute Diseases*. Chapter 3 points out the consequences of the current disagreements between doctors on the proper remedies to apply for particular complaints.

> The art as a whole gets a very bad name with ordinary people, so that it is thought that there is no art of medicine at all.' What one doctor prescribes as the best treatment, another considers to be bad. 'So that people say that the art is like DIVINATION. For diviners too think that the same bird is a good omen if it is on the left but is a bad one if it is on the right . . . while other diviners think the opposite.

Of course the doctors claimed that *their* signs were reliable, but diverse ones were used in different ancient medical traditions.

If divination can be said to be a mode of prediction, where does foresight fit into the taxonomy? Evidently it overlaps with prediction, but it does not necessarily involve a precise foretelling of what will come about. Rather what it does depend on is some sense of what is likely to occur, especially the probable consequences of different courses of action. We need it (don't we know) when we are dealing with questions of policy: what will happen if we deregulate the banks? That's not an ancient example, since they did not have much by way of banks. But how to deal with relations with foreign states, how to ensure the productivity of the land, even how to make taxation equitable; they all figure in our ancient sources often enough.

Many of the figures whom I have been citing were advisers. Certainly the Mesopotamian *ṭupšarru* were constantly advising their kings. Oracles were consulted in the Greco-Roman world to get advice on what to do: the Persians were about to invade Athens: Delphi said 'put your trust in your wooden walls' (there was a dispute about that one too: some said that meant the wooden fortifications of the Acropolis, but Themistocles said 'ships'. The ships interpretation won out and indeed the Athenians were victorious at Salamis). The oracle bones in China were consulted on many similar practical issues. Many of the great names in Chinese philosophy, from Confucius onwards, served as advisers: in his case he went from state to state never finding a ruler who was worthy enough to receive his advice. In fact a common generic name for those we call Chinese philosophers was *you shui*, literally 'itinerant advisers'.

The trouble with foresight is how to acquire it. How do you become good at it? Here there weren't any books from which you could learn the skill, like the divination handbooks that you were expected to learn by heart. There were, however, role models, figures of legend mostly. The Greeks had a paradigmatic foreseer who might be considered a pretty disastrous example to follow, Prometheus. One problem with Prometheus was that he had a younger brother, Epimetheus: he told Epimetheus never to trust Zeus, never to accept a gift from him. But when Zeus, disgusted that Prometheus had stolen fire and given it to human beings, took his revenge – that took the form of the first woman Pandora – do not let me get distracted – Epimetheus accepted it, and Pandora opened the box – or rather it was a jar – full of the 'gifts' of the gods, though these turned out to be every kind of evil, which were thus let loose on humans. Prometheus should have foreseen that it was not good enough just to warn his brother about Zeus. Then did Prometheus anticipate the punishment that Zeus would inflict on him (pretty vindictive this god)? Namely to chain him to a rock and have an eagle eat his liver; the liver however regenerated every night so that the eagle would be able to continue the torture as long as Prometheus stayed chained to the rock (see Figure 1.5). Of course in a sense if Prometheus had foreseen that, it makes him even more heroic and generous to humans. But we are not told he foresaw his own punishment and we might rather conclude that his foresight had its limitations. It generally does.

The Chinese had their role models, too, the Sage Kings, helpfully juxtaposed to the Legendary Tyrants, to get across the message that the former concentrated on the welfare of their people, of 'all under heaven' indeed, while the latter were self-indulgent and neglected their people, who suffered all sorts of misery as a consequence. But the historical figure Confucius also acted as a model. In one text he indicates where his priorities lie, and that is not with divination, though he does not reject divination entirely.

FIGURE 1.5 *Prometheus Bound* by Peter Paul Rubens

Zi Gong [one of his pupils] asks the Master: 'Does the Master also believe in milfoil divination?' To which the Master replies 'As for the *Changes*, I do indeed put its prayers and divinations last.' You can use the 'commendations' to get to the number, but 'if the commendations do not lead to the number, then one merely acts as a magician; if the number does not lead to virtue, then one merely acts as a scribe.' 'Perhaps it will be because of the *Changes* that knights of later generations will doubt me. I seek its virtue and nothing more ... The conduct of the gentleman's virtue is to seek blessings; that is why he sacrifices, but little; the righteousness of his humaneness is to seek auspiciousness; that is why he divines, but rarely'.

(Shaughnessy 1996, 241, modified)

Of course, understanding Confucius takes the most strenuous and concerted efforts of your reason and imagination. But there is a rather more accessible source of the kind of practical foresight you need: history. Two of the greatest historians, one Chinese, one Greek, were very conscious of the point. Sima Qian completed the work that his father Sima Tan had begun, the first great Chinese universal history, the *Shiji*, composed between *c.*100 and 80 BCE. This has several different sections, dynastic histories, chronological tables, treatises on such topics as agriculture and – yes – the study of the heavens (astronomy/astrology), as well as chapters devoted to the biographies of various types of people, including 'harsh officials', 'money-makers' and 'assassin-retainers'. Most of the chapters are punctuated with comments on the lessons that can be learnt from what has been recorded, introduced with 'The Grand Scribe says' (where sometimes this is Sima Qian, sometimes his father). Take the end of the story of the Rebel Xiang Yu (*Shiji* 7, 339.2ff.,). He had ignored the lessons of antiquity and tried to take the Empire by force. 'He called his task that of hegemonic kingship, but he intended to rule and control all under Heaven by force. But after five years he finally lost his kingdom and died at Dongcheng, but he still did not wake up and lay the responsibility for his mistakes on himself. Thus surely he was deluded to invoke the words: "Heaven is destroying me, and it is not anything I have done wrong in the use of the military."'

But it is not just the case that over and over again Sima Qian ponders the lessons of history: he makes it quite clear that his history itself is intended to be such a source of instruction. Twice we find texts that suggest that 'the past remembered is a guide to the future' (*Shiji* 6, 278.9ff., quoting Jia Yi) and that learning from the past can provide 'a key to success and failure in one's own age' (*Shiji* 18, 878.4ff.). That's a trope that antedates systematic historiography in both China and Greece. I am thinking of the recurrent moralising tales of how drink and indulgence in sex lead to you losing your kingdom, in China, and those relating to how pride, arrogance and hubris herald your eventual comeuppance in Greece. Foresight, the lesson is, can be gained by immersing yourself in the past. That is always a matter of interpretation, to be sure: but the outcome of different modes of behaviour, of different policies, was there for all to see.

And if the greatest early Chinese historian puts it that way, my Greek example, Thucydides, is perhaps even more explicit. He distances himself from his predecessors, whose histories were redolent of what Thucydides considered mere storytelling, the mythical indeed: he has Herodotus among others in his sights, but he names no names. 'The lack of the mythical may make my account less agreeable to listen to. But as for anyone who wishes to have a clear view both of the events that happened and those that will happen in such, or a similar, way in future, according to human probability, if they deem my account to be of use, that is sufficient for me.' What he offers, by contrast, is, in his famous phrase (Thucydides I ch. 22), not a 'competition piece for the moment', but a 'possession for always', a repository of wise advice for all to learn from. That is thanks to the assumption that human affairs repeat themselves, not in the sense of exact repetition, as in the eternal return I mentioned, but rather that there are constant principles at work.

It has sometimes not been noticed that Thucydides' account of the ravages of civil war in Greece is twice juxtaposed with his account of the plague that hit Athens in particular around the year 430 BCE. He himself suffered from it, and he tells us why he gives us a detailed description of it (Thucydides II ch. 48): 'I shall say what it was like, and set out the details from the study of which a person may be able to recognise it if it should ever attack again. [I shall do so] because I both suffered from the disease myself and saw other victims of it.' Whether there was just one pathogen responsible for every ailment 'the plague' covered may be doubted: attempts at retrospective diagnosis have varied and truth to tell all fail to account for all the signs and symptoms recorded. In equivalent texts in the Hippocratic Corpus relating to epidemics, it is more usually recognised that several diseases were present, even though certain conditions were particularly prominent. Neither Greek nor Chinese doctors could do much to help their patients when they suffered from acute conditions. Yet the intensity with which individual case histories and more general 'epidemics' were observed and recorded is witness to their valiant attempts to get a handle on pathology, and, as we saw in the text from

Prognostic, at least to be able to foresee likely outcomes. For successful treatment, ancient doctors needed more than just foresight – they needed effective cures – but so far as foresight was concerned, their dedication to the task of building up a database, and so to gaining just that ability to foresee outcomes, was extraordinary.

Thucydides thinks that human affairs are subject to factors that are more or less constant over time – just like diseases. One can go further. The account of the plague is there (so he tells us) so that we should be forewarned. His whole history (we may say) is similarly there to help us to be on our guard against what he represents as the sicknesses of the body politic, notably civil strife.

Foreseen Conclusions – and a New Point

So what lessons do I draw from the complex data that I have briefly surveyed? I feel that maybe I shall have failed in my task if my conclusions cannot be foreseen. So the summary of my argument will be brief, though I have one new point to add. Greeks, Babylonians, Chinese (and I could have added Indians and Egyptians too) were all obsessed, just as we are, with trying to be able to see what is in store, for individuals, for groups, for whole civilisations. As we have seen, in their attempts to do so, they tried out hundreds of different methods. Some were spot on: eclipses came to be predictable since their periodicities could be derived on the reliable basis of previous observations. Many more techniques were said to be based on experience but relied on non-existent causal connections – as some of the ancients themselves protested – though that did not stop them continuing to be used. Is that not so still today?

At the same time many ancient writers did their best to draw out the lessons from human experience: that was an important motive for historians, annalists, memorialists, writers of all sorts. This was not (usually) to make precise predictions, but to plot probable or possible consequences of different actions and policies. That was generally a matter of just seeing what followed from what, what had happened in similar circumstances in the past. But it was often a question of focussing on what is right, proper, beneficial; on values in other words. When that was the case, more was involved than just learning about events. It was not just a matter of prediction, nor just a matter of foreknowledge, but a matter of wisdom. It may have been in short supply in ancient civilisations, but how can we not agree the same is true today? It does not take a crystal ball to get the answer. Ancient foresight has lessons for *our* claims to foresight today, about our fallibility, our gullibility, our need to work hard to get it right and, above all, to do what is right – themes that are taken up in other chapters of this book.

Further Reading

Barton, T.S. (1994) *Ancient Astrology*. London and New York: Routledge

Bascom, W. (1969) *Ifa Divination*. Bloomigton: University of Indiana Press

Beard, M. (1986) 'Cicero and Divination: The Formation of a Latin Discourse', *Journal of Roman Studies* 76, 33–46

Bouché-Leclercq, A. (1879–82) *Histoire de la divination dans l'antiquité*, 4 vols. (Paris)

Djamouri, R. (1999) 'Ecriture et divination sous les Shang', *Extrême-Orient Extrême-Occident* 21, 11–35

Douglas, M. (1975) *Implicit Meanings*. London: Psychology Press

Denyer, N. (1985) 'The Case against Divination: An Examination of Cicero's *De Divinatione*', *Proceedings of the Cambridge Philological Society* NS 31: 1–10.

Evans-Pritchard, E.E. (1937) *Witchcraft, Oracles and Magic among the Azande*. Oxford: Clarendon Press

Harper, D.J. (1978–9) 'The Han Cosmic Board', *Early China* 4, 1–10

Hunger, H. (1992) *Astrological Reports to Assyrian Kings*.

Jones, A. (1999) 'Astronomical Papyri from Oxyrhynchus', *Memoirs of the American Philosophical Society*, 233

Kalinowski, M. (1991) *Cosmologie et divination dans la Chine ancienne*. École française de l'extrême Orient: Paris

Keightley, D.N. (1988) 'Shang Divination and Metaphysics', *Philosophy East and West* 38, 367–97

Moore, O.K. (1957) 'Divination – A New Perspective', *American Anthropologist* 59, 69–74.

Ngo Van Xuyet (1976) *Divination, magie et politique dans la Chine ancienne*. Paris: University of France

Park, G.K. (1963) 'Divination and Its Social Context', *Journal of the Royal Assyria* 8

Parpola, S. (1993) 'Letters from Assyrian and Babylonian Scholars', State Archives of Assyria 10

Rochberg, F. (2004) *The Heavenly Writing: Divination, Horoscopy and Astronomy in Mesopotamian Culture.* Cambridge: Cambridge University Press

Shaughnessy, E.L. (1996) *I Ching, the Classic of Changes.* New York: Ballantyne Books

Vernant, J.-P. (ed.) (1974) *Divination et rationalité.* Paris: Ed du Seuil

Schofield, M. (1986) 'Cicero for and against Divination', *Journal of Roman Studies* 76, 47–65

2 Foresight in Journalism

BRIDGET KENDALL

Why Journalists Are Wary of Foresight

We all watch or listen to the news and read the papers not only to find out what has taken place, but to get a steer on what might be coming up. Journalists may like to think they deal with hindsight or possibly insight, but if they can also show foresight, undoubtedly they serve their audience better.

However there is an irony here. Journalists may be asked for their views on what might happen next, but many of them do not relish the prospect. Trainee reporters are taught to observe and probe in order to give a faithful account of what has been seen and heard, but not to stray into speculation. Even specialist journalists who focus on a particular subject are generally wary of acting as seers. They prefer to back up their analysis with known facts, attributed quotations, and the solid rock of proven experience. It is the past, not the future, which is the foundation stone of their credibility.

Commentators and columnists may be encouraged to give personal views and venture predictions. But journalists dealing in daily news want their accounts to be trustworthy and reliable. So when it comes to projections they may well hedge their bets by sticking to the obvious, offering a choice of scenarios, inserting caveats, or by using a formulation which is vague or hard to disentangle, so that the risks of being proved wrong are kept to the minimum.

Trust and credibility are highly prized not only by individual journalists but by their news organizations as a whole. If the audience does not trust the source or conduit of information, they are less likely to turn to it on future occasions. One of the easiest ways to damage a reputation for reliability is to make a rash prediction that turns out to be wrong. Indeed those who make a habit of offering bold predictions run the risk of being lampooned.

The late and illustrious newspaper editor and columnist Lord Rees-Mogg, was a highly influential and well-informed journalist. He was also not shy about making predictions. Indeed he was so categorical and sufficiently often off the mark, that the satirical magazine *Private Eye* dubbed him 'Mystic Mogg', and when he died in December 2012 it ran a full-page spread as a tribute, outlining some of his predictions:

These included:

1   in 1991 the Western world would enter 'a decade of escalating economic and political disorder, unparalleled since the 1930s';
2   in 1996 Colin Powell would be elected America's first black president by a large majority;
3   in 1997 it would be strange if Labour won the British general election by a landslide;
4   in 2011 the euro crisis was already over.

Not all these predictions were wrong. One could argue that in foreseeing a decade of economic and political disorder, or America's first black president, Lord Rees-Mogg was simply ahead of his time and that his error was not of judgement but of timing. Besides which, accuracy of predictions is not everything. Lord Rees-Mogg was always entertaining and knew it. 'It's not my job to be right,' he is quoted as having said. 'It's my job to be interesting.'[1]

He had a point. There is almost nothing worse than a boring piece of journalism: however accurate it is, it will find few readers.

Of course a reputation may also be tarnished by being *too* cautious or by failing to anticipate momentous developments which land right in front of you. This is where journalists often fall short of foresight. At the start of the election campaign ahead of the 2008 US presidential election, how many journalists predicted that the United States would be about to elect its first black president? Going further back, how many of us (myself included) stationed in Moscow at the time, anticipated that the Soviet Union would collapse quite so suddenly at the end of 1991?

More recently, how many Western journalists – especially those who were not based in Cairo or Tunis – saw the Arab Spring coming? And how many recognised that the waves of protests which toppled presidents in Tunisia and Egypt at the start of 2011 would lead to revolutions in Libya and Syria?

I know I did not. I remember a radio broadcast from February 2011 in which I argued that protests similar to those which brought down presidents in Tunisia and Egypt were surely unlikely to shake the regimes of Libya and Syria, because Libya's leader, Colonel Muammar Gaddafi and Syria's president, Bashar al-Assad, would learn lessons from what had happened and take measures to impose a stricter grip on society to prevent mass unrest. Less than two weeks later, I found myself reporting on the Benghazi uprising in Eastern Libya which was to mark the start of the end of Colonel Gaddafi's rule.

Being a journalist means walking a tightrope. It is a perpetual balancing act between risk and caution: trying to be sufficiently agile to seek out and push the stories that matter, yet not going out on a limb only to be proved wrong. It is the tension between seeking safety in numbers to hunt as a pack, and striking out on your own as a maverick, a lone wolf.

It is also the tension between being first and being right. Some news organisations place a greater premium on being first, getting the scoop to splash on their front page or beating other broadcasters with 'Breaking News' which cuts into scheduled programming. Others, including the BBC, believe their reputation rests more on being accurate, on checking out new information from multiple sources before airing it, even if this means they may sometimes lag behind the competition. After all, one hasty assumption or off-the-cuff remark can come back to haunt you and colour your reputation for years to come. And these days when everything is always archived somewhere online, a mistake may well be in the public domain forever.

Perhaps the most famous wrong call to be aired on the BBC was not strictly from a journalist, but from someone whose face was very familiar to TV news audiences (Figure 2.1).

The veteran weather forecaster Michael Fish will forever be remembered for one throwaway line in a BBC lunchtime forecast in October 1987, when he reassured those who feared a big storm was coming that there was nothing to worry about:

FIGURE 2.1   The Famous Lunchtime Broadcast by BBC Weather
Forecaster Michael Fish on 15 October 1987

> Earlier on today, apparently, a woman rang the BBC and said that she
> heard that there is a hurricane on the way ... Well if you're watching,
> don't worry. There isn't. But having said that, actually the weather will
> become very windy. But most of the strong winds will be down over Spain
> and across into France.[2]

Fateful words ... a few hours later the worst storm to hit South East
England for several centuries was to cause record damage and killed eighteen
people. Michael Fish was never allowed to forget it. To his credit, he cheer-
fully acknowledged he had been wrong. It even won him a slot in a video
montage during the opening ceremony for the London 2012 Olympics. It is
a lesson that if your prediction is outright wrong, it can be almost impossible
to live it down. You might well need to embrace it as part of your legacy.

Tools Which Help with Foresight

So what are the tools of the trade which may help journalists do better at
prediction and foresight?

I want to discuss three of them: two are relatively new (social networks and large data sets), while one is as old as journalism is (access to privileged sources).

Social Networks

Social networks, that compendium of interactive communications tools, extend a journalist's scope to a wider group of actors and opinions. Facebook, Skype, and Twitter have opened up new possibilities for all sorts of groups to leap over old boundaries and reach each other in virtual space, while remaining invisible below the radar of official scrutiny. In the old world of mass communications, professional journalists were at the heart of the information gathering and dissemination – the world of newsroom hubs, expensive satellite link-ups, and news delivered to desk-tops via wire services. Now social network tools have revolutionised and decentralised ways of sharing information.

One correspondent to take an early interest in the power of social networks was the BBC's Chief International Correspondent, Lyse Doucet.

Lyse Doucet was an early convert to using social network sources. She has said that she first became aware of the potentially transforming effect of social networks for journalism in 2009, during the so-called Green Revolution in Iran. When word came of the first street protests, older journalists in the BBC Tehran Bureau rushed to their desktop computers to log on and try to work out what was happening by looking at news websites. At the other end of the office, younger journalists were on their phones, checking texts, blogs and – the latest new thing – twitter feeds. Their sources were faster and more detailed – if unverified: the latest news straight from the street.[3]

By the time Lyse Doucet came to cover the Arab Spring revolutions in 2011, twitter had become her touchstone, a means of keeping abreast of fast-moving events, of being tipped off about possible new developments, and a way of being both challenged and supported by the audience, as she described in a recent lecture.

> January 25th, Cairo. We were on the top floor of the BBC Arabic Service's broadcast centre in Agouza. Even on the 19th floor you could taste and smell the tear gas. It was unfolding in the streets below.

> I was interviewing an old friend and former colleague, Yosri Fouda, who is now, arguably, Egypt's most prominent and influential journalist. And Yosri says on the BBC: 'Look at the tear gas below! Those tear gas canisters are made in America!'
>
> When I checked the Twitter feed, someone had sent me a message, saying: 'Lyse Doucet! Shame on you! You didn't challenge Yosri Fouda when he said those canisters are made in America!'
>
> I thought 'Oh dear . . . '
>
> But then within seconds another tweet arrives. Someone took a photograph on the streets, and there it was – a close up of the canister and it said 'Made in America.'
>
> And I said, 'Thank you Twitter'[4]

Lyse was an early convert.

For me the revelation came in February 2011 during the uprising in Benghazi in Eastern Libya, the start of the Libyan revolution which was eventually to depose the over-forty-year-old Gaddafi regime in Libya. At the outset the Libyan leader, Colonel Gaddafi, attempted to impose a media blackout but it did not work. Information emerged anyway, through texts, tweets, YouTube footage, and Skype interviews corralled by a global network of activists. The same pattern repeated itself in Syria a month or so later. President Assad's troops tried unsuccessfully to isolate Homs and cut off protestors and rebel fighters there by disconnecting government-controlled communication links.

These uprisings have shown that in many of today's conflicts decentralised communication networks can often provide ways to circumvent official censorship or blackouts. The same social networks provide a valuable resource for journalists the world over, allowing them to tap into new sources of information, reach a wider spectrum of opinion, and interact with both the subjects of their reports and the audiences who consume them.

It is a resource which needs to be used with caution. Social networks can mislead as well as enlighten. YouTube footage can be doctored. Information encapsulated in brief messages on twitter can turn out to be deliberate rumours, dressed up as information. It all represents an alarming mountain of new material which needs to be processed carefully.

It will never replace the time invested in actually going to a place, to meet and listen to people face to face. But it is a new way to enhance a journalist's grasp of what is happening – and what might be round the corner.

Data

Another useful new tool for journalists lies in the increasingly sophisticated use of data. As Michael Fish pointed out in a recent BBC interview, given the amount of computer data now available to meteorologists, weather forecasts have become much more detailed and reliable. These days it would be highly unusual not to see a hurricane coming. In 1987 it was the limitations of technology, not his forecasting skills, which was the problem:

> The computer of course is the thing that did the forecast, but you can't blame it (. . .) It was unfortunate that the computer lacked a huge amount of data from the area where the storm was developing. (. . .)
>
> Of course we didn't have the mega computers that we've got now, and the forecasts have got so much more accurate. (. . .)They have got more and more accurate. You can actually say that a three day forecast is better than a 24 hour or 36 hour forecast used to be a few years ago.[5]

Improved data has not only helped meteorologists. The volume of easily accessible public data and the new software tools which can interrogate it have created a new field, known as data journalism. The primary aim is to extract patterns and then turn the sometimes surprising insights hidden in large data sets into journalistic stories, written up in text, or explained through graphic visualisations. It is an especially useful tool when making projections. In 2012 one American practitioner hit the headlines through his spot-on analysis of the likely outcome, state by state, of the last US presidential election.

Nate Silver is a *New York Times* blogger and author who combined polling data to come up with remarkably accurate picture of how Barack Obama would sweep to victory for a second term as US president.

During the election campaign Nate Silver's prediction was controversial because it was at variance with the received wisdom of many pollsters. Their findings suggested that the race between President Obama and

Governor Romney was too close to call. Many polls in swing states showed that the two candidates were neck and neck, hovering within just a few percentage points from each other. But for weeks before the election, Nate Silver was insisting that the race was not narrow at all and Barack Obama would win comfortably.

When he was proved right, he was promoted to superstar status. He was dubbed the 'blogger and geek' who had 'nailed the answer in 50 states' and 'Lord and God of the Algorithm'. Andrea Canning for NBC's *Today* programme summed up his winning formula:

> Avoiding the spin of talking heads, Silver sticks to numbers, averaging state-wide polls, factoring in for uncertainty and then running a variety of scenarios for how these probabilities would play out.[6]

Access

However, data and numbers are not everything. Old-fashioned journalism means sifting through all sorts of information, not all of it quantifiable. And a third resource which journalists often take for granted is their privileged access to events and people at all levels, both at the very top of society and at the very bottom.

In our era of 'citizen journalists' and bloggers, it is this access which perhaps most distinguishes professional journalists from other analysts and observers. Even if sometimes the information which governments and other official sources are prepared to impart is so guarded and bland it hardly seems worth having, it is always valuable.

Inside access can be seductive. A journalist has to be careful not to fall into the trap of swallowing wholesale confidential information. He or she has to weigh up what they are told against other evidence. But this is a chance to gain a unique insight into the thinking of leaders whose decisions may be important – a point of comparison which helps illuminate the way forward.

Insights can be gleaned from briefings with officials on government policy, face-to-face public interviews with politicians, and private conversations with inside contacts. These off-the-record briefings may help confirm hunches or modify assumptions. They may also expose unrealistic ambitions or downright hypocrisy in a government's policy when compared to what is happening on the ground.

Both public rhetoric and more private assessments may fall short when tested against raw reality. Diplomats, politicians, and military generals are more likely to brief journalists on their hopes and aspirations for a policy or military strategy, rather than chew over the possibility of failure. But get on a plane and go to a front line or a failing state or a country in the midst of a revolution, and match those political aspirations against what is really happening. It may be a very different picture.

Sometimes on-the-ground access may allow you to foresee how it will all go wrong. Sometimes it may reveal that a government policy conceived in a capital many miles away is making a positive difference. But sometimes you may find, counter-intuitively, that a conflict is not all as serious as it is made out to be.

Moldova's War 1990

In the early 1990s when I was stationed in Moscow for the BBC, I once took a trip to Moldova (or Moldavia as it was still then called) a tiny Soviet republic on the USSR's border with Romania, where simmering tensions between Moldova's Romanian speakers and Russian speakers had erupted into what was being described in Soviet newspapers as a mini civil war.

Many Romanian speakers, along with many other ethnic minorities in the republics which made up the Soviet Union, were beginning to chafe at Moscow rule and talk of the advantages of independence. But the heavily industrialised and militarised Russian-speaking region of Trans-Dniester inside Moldova was having none of it. The Russians there wanted to remain part of the Soviet Union. From Moscow there had been alarming reports of sporadic clashes between opposing sides which, according to Soviet officials, had the potential to turn into a full-blown bloody conflict.

It was not easy to get to Trans-Dniester from Moscow. It required a flight to the Black Sea port city of Odessa and a long and jolting taxi ride. On arrival I found a sleepy backwater where nothing much seemed to be happening. At a school turned into a temporary military headquarters, officials on the Russian-speaking side refused to accompany me to the front line. Instead they gave me a rather unconvincing white strip of cloth to tie to my upper sleeve to show I was a member of the media rather than a combatant and pointed me in the right direction.

I walked past one bored Russian soldier manning a checkpoint surrounded by sandbags. I passed two more empty checkpoints and several abandoned military vehicles. The road twisted and turned a little, going past fields and hedges. And then I came across a new group of soldiers eating sandwiches. They looked at me in surprise but no alarm. It turned out they were Romanian speaking and I had walked across the so-called front line and had not even noticed it.

It is true that this so-called 'frozen conflict' was to remain a burden on Moldova for years to come. But it was not the alarming war I had read about in Moscow. I concluded that President Gorbachev's beleaguered Soviet government wanted to play up the dangers of civil strife in Soviet republics out of desperation, in order to make the case for keeping a strong central power.

Usually politicians want to play down calamities, not play them up. But this was an unusual case of the reality being less dangerous than official rhetoric suggested. And only by being able to go to Moldova to see the stand-off for myself was I able to gain that understanding.

Interviewing Vladimir Putin

Sometimes access to the top can give you useful insight into the character and likely behaviour of world leaders. But whether it makes it easier to foresee their future behaviour is another matter.

In March 2001, I was invited to Moscow to take part in an interview with the then still-new Russian president, Vladimir Putin. I was to put questions to him on behalf of the BBC's global audience, alongside two Russian journalists who fielded questions from Russian viewers.

At the time President Putin had been less than a year in office and had not given many interviews. I think the Kremlin thought it would make him look flexible and modern to be seen to be communicating with people via the internet.

They had built a special TV studio inside the Kremlin, complete with flashing neon lights and an electronic backdrop to show off the many Russian questions for the President flooding in from across the country. It was a surreal experience. But it was also an extraordinary chance to gauge what Mr Putin was like close up.

Perhaps my strongest impression was that in 2001 there was not one Mr Putin but two. The first was quite hesitant and even shy, not quite sure how to answer when, for example, he was asked personal questions about his favourite books and music, and about his family life. The other Mr Putin was quite different: fierce, assertive, and especially touchy on human rights issues, or when asked challenging questions about security and terrorism.

I put one question to him on behalf of a Danish listener who wanted to know if President Putin thought Chechens would ever consider Russia an ally, given all the destruction and 'cruel methods', as she put it, used in the recent Russian military attack on them.

Clearly irritated, Mr Putin thanked her for the opportunity to correct the mistaken views of so many Westerners about the Caucasus. His reply was that the Russian army was not waging a campaign against the people of Chechnya, it was responding to terrorism and the Chechens understood and appreciated this – whereupon I challenged him, saying that I had just come back from visiting Chechen refugee camps and found the general attitude towards Russians was very negative. He visibly bristled.

Nowadays political analysts of Russia often refer in shorthand to the Putin presidency as a two- or possibly three-stage process. Putin One, in his first term, was prepared to reach out to the West, repair relations with NATO which were damaged by the 1999 Kosovo bombing campaign, invite Western investors in, and offer close collaboration on fighting international terrorism.

Putin Two took over in 2003 or 2004, around the start of his second term. He sacked his reformist prime minister; he oversaw or at least did nothing to stop a legal campaign against the oil tycoon Mikhail Khodorkovsky which removed him as a potential rival by sending him to a Siberian prison camp; and – influenced probably by Ukraine's Orange Revolution – he began to voice his fears that foreign agents (from the Islamic world and from the West) were seeking to manipulate or destroy Russian power from within by funding charities and other non-governmental organisations.

Putin Three refers to the President's even greater intolerance of those he sees as his critics and detractors since he was re-elected president in March 2013.

So given the chance to meet Vladimir Putin face to face in 2001, did I see the change from Putin One to Putin Two coming? Did that early interview give me foresight into what sort of Russian leader Vladimir Putin would become?

Well, yes – and no.

In private, when asked what I made of Mr Putin, I would frequently say that I thought his ability to switch between two personae was unsettling. The more-personable Mr Putin seemed flexible, even a little insecure, and open to contact with foreigners. But especially when it came to matters of state and security, his thin skinned refusal to accept any criticism was worrying. It veered on what looked like paranoia. And the Mr Putin with the piercing blue eyes and turned down lip was not someone you would want to antagonise.

Only in hindsight did I realise that my impressions were a useful hint of what was to become his increasingly authoritarian manner and style of leadership. When I met him again for a second interview in 2006, the informal, more-flexible Mr Putin was no longer in evidence. Putin Two had apparently taken over. But in 2001 I would not have felt confident to make that prediction in public.[7]

Even now, the implication of the question (which way might Mr Putin turn, towards democracy or dictatorship?) serves as a reminder that often an either/or option can be too limited for a useful prediction. Perhaps journalists should not feel too apologetic about offering a range of options. Maybe journalistic foresight does better when it acknowledges the need for uncertainty.

Putin's Russia has proved itself to be neither a Western style liberal democracy nor a full-blown dictatorship, and the leniency or harshness of the government's responses fluctuates, depending on many factors.

Unlike the Soviet Union, Russia's current government does not ban free speech nor forbid contested elections. It does not bar Russian citizens from travelling abroad. It has been, in the Kremlin's own parlance, 'managed democracy'. Criticism of the government is allowed to flow freely on the internet or in small circulation newspapers, but on the television channels beamed into every Russian home it is virtually non-existent. Opposition candidates can run in election, but usually only if officially sanctioned. Anti-government protests are on occasion tolerated, but intimidation through a heavy police presence and the regular arrest of organisers serves to discourage high turnouts. And those who continue to defy the authorities have a curious habit of finding themselves investigated for corruption.

Perhaps when it comes to foresight, not only predicting a range of options, but making a stab at gauging the probability of their coming true on a scale of nought to ten would be more useful than offering a categorical black or white assessment. If an unexpected outcome occurs, after all, such a scale would then usefully signal how extraordinary it was.

Depth or Distance

Is it true that the more you know about something, the easier it is to have foresight of what will happen? Or is there a danger of not being able to see the wood for the trees? Or of being paralysed by seeing multiple options, cursed by indecision in the face of too much information?

If foresight is about wisdom, the ability to identify an underlying trend or shift and recognize its importance, then it would seem logical that the more context and understanding the better.

In a straw poll of some of my BBC colleagues, I found most agreed that even though a fresh pair of eyes on a situation can be illuminating, generally the more you know about something, the better your judgement is likely to be.

As one veteran BBC journalist once said to me, 'The knowledge you reveal on air should be the tip of the iceberg.' In other words, all judgements should be based on a mountain of hidden information and experience which you do not necessarily display but which informs the selection of every adjective, quotation, or shift of emphasis.

But there are pitfalls that come with this.

Journalists who are especially good at their job, able to sift through data and use their access, experience, and intuition to piece together a puzzle may produce the scoop of the decade. But they will not always be congratulated by all on their foresight. Sometimes they may find themselves accused of being too close to their sources and benefitting from special favours.

On my old patch, the former Soviet Union, this was a particular hazard. When a foreign correspondent had good connections and a good understanding of what was going on, it was all too common to hear suspicions aired about possible links to the security services. One American journalist of Yugoslav extraction who fell victim to this was a man called Dusko Doder. He was Moscow Bureau chief for the *Washington Post* in early 1980s, spoke good Russian, had good contacts, and enjoyed a reputation for ground-breaking reports on Soviet politics.

As he subsequently described in his book *Shadows and Whispers*, in 1984 he had a world exclusive. Ahead of anyone else he reported the death of the Soviet leader Yury Andropov. This was not thanks to a contact in the KGB or Communist Party having tipped him off, but because he was expert at picking up clues of the ailing Soviet leader's likely demise, nuances, and details.[8]

He noted a speech on television which omitted the mandatory homage to the General Secretary. He realised a scheduled jazz concert on the radio had been replaced by sombre music. He noticed a press report which revealed Andropov's son had returned early from abroad. And finally on a late-night walk through the city centre, he spotted the tell-tale lights on in the KGB headquarters and the Ministry of Defence, usually a sign that something momentous had happened. All these hints matched what he had observed when the previous Soviet leader, Leonid Brezhnev, died fifteen months earlier.

Unsurprisingly, none of his official contacts would confirm his suspicion. Getting information out of Soviet officials was always difficult and this was the most sensitive subject of all – the death of a Soviet leader. But the hints and hunches he had amassed were enough to convince him that he had enough material to file a story which would be factually correct, but which, by laying out all the evidence, would offer the conclusion that the Soviet leader, Yury Andropov, must be dead.

Unfortunately for him, in Washington that night his editor was attending a gala dinner at the US Department of State and took the opportunity to run the story past both the American Under-Secretary of State (Lawrence Eagleburger) and the Soviet Ambassador (Anatoly Dobrynin). Both of them dismissed the scoop and suggested that Dusko Doder's imagination must have got the better of him. So instead of being a *Washington Post* front-page exclusive, his report was toned down and published on page 28.

The next morning when he woke up in Moscow expecting to find his byline on the front page, he was in for a disappointment. To make matters worse, later that day his foresight was proved right: the Kremlin confirmed that Yury Andropov had indeed died and there would now be a new Soviet leader.

This cautionary tale does not end there, however. Eight years later, an article in *Time* magazine raised the possibility that Dusko Doder might have been co-opted by the KGB while he was in Moscow and accepted KGB money.

Dusko Doder was outraged. He took *Time* magazine to court for libel to clear his name, arguing that his exclusives from Moscow had not been the result of special favours, but good reporting. The case was settled out of court. *Time* magazine paid his legal fees and apologised.[9]

Subsequently, Dudko Doder said that he felt he had been vindicated. But it is a reminder that you can find a shadow cast over your reputation not only for getting something wrong, but for getting something too right. In other words, you can pay a price for passing the test of foresight.

Nowadays it is often financial journalists who find themselves the subject of scrutiny.

The BBC's business editor, Robert Peston, has won a string of awards for his exclusive reports, including most notably his scoop in September 2007 that the bank Northern Rock was seeking emergency financial help from the Bank of England.[10] Before long there was a run on the bank and it looked as though Northern Rock might collapse.

The role of the media in a financial crisis became a hot topic, both in the press and in parliament. The question being mulled over was this: if a journalist breaks a story about, say, a bank which suggests it is in serious trouble, are they making it more likely that the bank could go under? And should this journalist be applauded for showing extraordinary foresight as the result of diligent work? Or should they instead bear some responsibility for shaping events as well as anticipating them? And indeed, might it be that wittingly or unwittingly they were acting on information which served someone else's interests?

At the House of Commons, the Treasury Select Committee invited Robert Peston and other financial journalists to answer questions about their journalistic practices. When asked by MP Michael Fallon if one of the reasons for his success was particularly close contacts and access to the Treasury, Robert Peston seemed to relish the chance to defend himself:

> You will not be surprised that the one area where I am uncomfortable talking in public is over sourcing of any sort. Over the years I have benefited from private conversations with members of this committee – and I think it would be very unlikely that many of those members would wish me to divulge those sorts of chats.
>
> The one thing I would say is I have been a journalist for 25 years. I've done political journalism. I've done business journalism. And I like to think I have decent contacts in the city, in government, in Whitehall, in the city and with regulators.
>
> When I do a story it is normally a process of putting together a jigsaw puzzle. Very, very, very rarely in my life has someone rung up and said 'I've got a corker – here you are', and handed me something on a plate. It almost never happens.
>
> I know a bit about banking. I've been a banking editor in the past on the *Financial Times*. In the summer of 2007 when markets closed down, I concluded that this was likely to be the biggest story of my career and I immersed myself in it and more or less everything since then has been a process of talking to probably hundreds of people over that period, working out the trends and working out what the stories were.[11]

Foresight: Cause and Effect

It is not just financial journalists who may have to think carefully about the 'cause and effect' impact of a report which could be construed as a forecast.

In war, or indeed during any emergency, journalists also have to be aware that their reports, which nowadays will reach a global audience in minutes, could affect the outcome of the very story they are telling.

Once British journalists were observers whose reports might take hours or even days to get back to London, and whose work was largely seen and heard by an audience sitting in their living rooms many miles from the scene. Now any report from a conflict zone reverberates back into the place it was coming from. It can become real time intelligence which may put those being reported on in danger: an ethical dilemma.

In describing a chaotic hostage siege, like the dramatic kidnapping of workers at a remote desert gas plant in Southern Algeria in January 2013, journalists have to weigh carefully whether it is wise to reveal facts about the captives' likely whereabouts. If it is known that some have escaped and hidden elsewhere in the compound, is it advisable to air this information, in case the kidnappers on site learn of it and decide to look for them?

If a hotel is taken over by gunmen, as in the case of the siege of the Taj Mahal hotel in Mumbai in 2008, do you really want to broadcast the fact that you have been in mobile phone contact with guests locked in their bedrooms? What if your report prompts gunmen roaming through the hotel to make a more thorough search of the building?

If you are crouched in a bombed out house in a suburb of Homs during the Syrian conflict, like the late *Sunday Times* correspondent Marie Colvin and her colleagues in 2012, should you steer clear of giving live interviews by satellite from your location? It is thought that it may have been the Syrian military picking up her signal and launching an attack on the house where she was staying which led to her death.

And when it comes to covering natural or man-made disasters, journalists have to be equally careful. In 1984, the BBC correspondent Michael Buerk and his camera crew went to Ethiopia to cover the worst famine there for a century. In his report he described it as 'a biblical famine in the 20th century' and 'the closest thing to hell on Earth'.[12]

The reporting on that famine by Michael Buerk and other colleagues shocked people in Britain and helped bring world attention to the disaster. It was an early example which showed just how powerful this sort of journalism can be.

Since then, increasingly, aid agencies and other development organisations who want to prevent new disasters have become highly attuned to the ways they can harness this sort of journalism to get their message across. They know that politicians take note of powerful reports on the main broadcasting outlets and on the front pages of major newspapers. One effective way to get governments to act is to wage a campaign through the media to force public attention on to an issue and put it at the top of a government's agenda. So how far should a journalist go in forecasting a disaster?

Most correspondents would say they apply the same criteria to this sort of story as they would to any other piece of journalism: reporting what they see, what they hear, and what they find to be credible – and what they have found out for themselves at first hand by visiting the area in question. Indeed those who cover these sorts of disasters on a regular basis are probably especially careful not to be seen as advocates for campaigning organisations.

But on occasion correspondents who have flagged up possible famines in various parts of Africa have been accused of exaggeration. There have even been allegations that they played up emergencies precisely to make a splash on a front page or a prominent TV report. And if you warn of an emergency which does not materialise, the impact can of course be counterproductive. Not only does it undermine you own reputation; crying wolf may mean that next time a real plea for help gets less attention.

But surely there is also an argument to be made for going to the scene of a potential disaster, precisely to tell the world that an emergency is imminent, but could be averted if swift action is taken?

In 2011, aid agencies feared a new food emergency in the Horn of Africa. There was not yet a famine, but they were worried that without speedy assistance from abroad it soon could be. The BBC decided this was news worth reporting, even though it was only highlighting the possibility, not covering the fact of a famine. Reporters and TV presenters were sent to the region to gather material, which was played out across TV and radio. The aim was to draw awareness to a prediction, in the hope it would not then come true. One of the journalists sent to Northern Kenya was the BBC's Ben Brown. Like his colleagues, he was careful to make the point that the disaster was caused by drought and civil war, but could not yet be classed as a famine:

> The United Nations says this is not a famine yet, but it could be. At the moment they are classifying it as a humanitarian emergency – a situation which they say is rapidly deteriorating. It hasn't rained properly around this region for two years running and these people are facing their worst drought for decades.[13]

It's an interesting new development in the complex relationship between journalists, aid agencies, and government – a much more sophisticated collaboration. In this instance all sides 'laid bare', if you like, the propaganda purpose of the reporting, and relied on the public to appreciate that this did not detract from its value.

The Limits of Logic and of Lessons from History

The opposite of foresight is hindsight. The modest ambition of any journalist is to be able to look back at their work and conclude that there is nothing they have reported which they would now like to retract or amend. Sadly that is not always how it is. In the huge volume of reports which journalists churn out, there is often cause for regret.

Perhaps the most infamous example of this is how the mainstream Western media reported the run up to the 2003 US-led invasion of Iraq.

There has been plenty of criticism for the way some Western journalists went with the flow of government thinking when it came to the crucial question in 2002 and 2003 of whether Saddam Hussein, the leader of Iraq, might be hiding weapons of mass destruction. Some media outlets and journalists have since admitted they were too credulous of British and American government claims that those weapons of mass destruction existed and posed an international threat.

Perhaps one of the main errors of judgement was to place too much weight on past evidence and on logical explanations – an easy trap to fall into, when journalists set such store by facts and rational argument. It was well known that Saddam Hussein had developed a chemical weapons programme. He had used chemical weapons in the 1980s, against Iranian soldiers in the Iran/Iraq War, and against his own people, most notably in a notorious attack on Iraqi Kurds in 1988. After the 1991 war in the Gulf, UN weapons inspectors had found substantial stockpiles of chemical munitions in Iraq and destroyed them.

And in the wake of that first Gulf war it also emerged that Saddam had created a large and clandestine biological weapons programme. Though by 2003 he was claiming it had been destroyed: it was hard to be sure that he was telling the truth.

The logical assumption, given his past behaviour, was that a leader such as Saddam Hussein might well be hiding illicit weapons, so that he could turn to them – as he had done before – as a weapon of last resort.

That was why it seemed reasonable in 2002 to send in more UN inspectors to look for any remaining stockpiles or other evidence of a weapons programme. His apparent failure to collaborate fully with them only intensified suspicions that he had something to hide.

All these were rational assumptions. But in 2005, two years after the US-led invasion, it transpired that logic and the lessons of history do not always lead to the right conclusion.

After exhaustive searches in the now-occupied Iraq, the Iraq Survey Group of international weapons inspectors finally concluded that it was unlikely Saddam Hussein had weapons of mass destruction. Their belief was that he probably *had* destroyed his biological stocks in the mid-1990s as he claimed, and although he may have wanted to keep the capability to restore his chemical arsenal in due time, in 2003 there was no immediate threat.

So why, then, did he not help himself by being completely open with UN inspectors, in order to give them a reason to clear him of suspicion? The answer, the Iraq Survey Group argued, was that Saddam Hussein hoped to have it both ways. He may have wanted to avoid a US invasion but he also wanted to keep his image of the strong man of the Middle East, possibly to impress his underlings, or maybe to keep potentially dangerous rivals in the region at bay.

In other words, above all, Saddam Hussein desired to be respected and feared. So even though it was not in his own interests, he could not bring himself to deny outright the possibility that he could still wield weapons which gave him an aura of absolute power.[14]

In some contorted way, this crazy mindset made sense. But could we have foreseen it?

The example of the Iraq War is an interesting lesson. If the Iraq Survey Team was correct, what Saddam Hussein was doing was not entirely rational. But sometimes an outcome cannot be explained by logic, or tracked on the basis of experience, or arrived at by looking at historical precedent. Human beings are not always rational animals.

Their behaviour and their choices can be quirky. Nate Silver's algorithms are all very well, but sometimes relying on data or past practice can be a trap.

For example, in the United States it used to be gospel among election reporters that whoever made it to the presidency would first have to carry the New Hampshire primary for their party. Then in 1992 Bill Clinton broke the mould – he failed to win New Hampshire and nearly ducked out of the race, but – ever the comeback kid who never gave up – he went on to win the election nonetheless.

Another piece of conventional wisdom was that when it came to electing presidents, the American people rarely voted for senators. Instead they favoured those who had already wielded executive power as governors. It was true in the case of Jimmy Carter, of Ronald Reagan, Bill Clinton and George W. Bush – all of them governors who went on to become president. For decades the formula held true: the last senator to move directly from the Senate to the White House was J. F. Kennedy back in 1961.

But in 2008 the American people elected not just a first-time senator with one of the most liberal voting records in the Senate, but a black man as well. What use was the old adage about senators and presidents now?

Perhaps it should not have been so surprising. The United States is an extraordinarily dynamic place, constantly reinventing itself. Its people are instinctively drawn to innovation, including new ways of behaving and voting. For all its conservatism with a small 'c', one of the United States' strengths is that it relishes breaking new ground and challenging stereotypes.

But there is a broader point here.

Truly impressive foresight is to be able to see ahead to those really big changes which will have far-reaching consequences – the paradigm shifts which not only overturn election lore, but which change the world on its axis. And do journalists excel at this?

I would argue that often they do not, precisely because they seek to back up assumptions with reasoned and fact-based arguments. When you cast your mind forward to a future which has not happened yet, you need something more – a leap of imagination or an insight which comes from being able to step back from the self-evident and expected, to put it in a different framework and imagine a corner turn which history could not have foreseen.

And sometimes it is the new observer who brings insights from another discipline who is best placed to see a crisis coming.

Gillian Tett is an award-winning journalist and commentator for the *Financial Times*, the paper's former US managing editor and one of the paper's stars. Before the financial crisis of 2008 she was a highly respected journalist but hardly a household name. The crisis made her famous. She was one of the few city journalists in London who as early as 2006 foresaw the possibility of a crisis in global credit markets and warned of the dangers of over-priced assets in her *FT* articles.

In 2006 she had returned to London from heading the paper's Tokyo Bureau to work on the City desk and write about the rarefied and highly technical world of collateralised debt obligation and credit default swaps. She has admitted it was not at first a subject she was a specialist in. But, unusually for a financial journalist, Gillian Tett was also a social anthropologist, with a PhD from Cambridge University, whose thesis was on the subject of marriage rituals in Tadjikistani villages.

She soon realised that her training as an anthropologist gave her a useful tool to observe the tribal behaviour of bankers in the world of high finance. It was helpful both because anthropologists look at cultures holistically, from the point of view of people and society, not just through the narrow prism of finance; and because it was a discipline which allowed her to analyse bankers and what they did from an entirely fresh perspective. She soon identified a worrying and dangerous flaw in their financial model, as she described in a keynote speech she gave at a conference for the Royal Anthropological Society:

> As I sat there and listened to their founding mythology about how credit derivatives were supposed to work, I could see that there were some very big intellectual contradictions.
>
> For example, one of the great founding myths about credit derivatives was that the bankers were creating the perfect free market in credit, where anything could be traded and everything could be priced according to free market principles. That was the great theory.

In practice when you look back at what was going on in 2005 and 2006, many of the products were so unbelievably complicated that they were impossible to trade.

And so actually, far from having a free market in these new credit products, you had an entirely fictional market. Products were priced on the basis of computer models, not free market prices – because they were too complicated to trade.

So there was a very fundamental intellectual contradiction, which almost no one could see back then, because they had all 'drunk the cool aid' – they believed in their ideology.

Not because they were bad people, let me stress. It is very convenient to blame the bankers for being greedy or evil. I don't think they were. They wanted to make money and they believed their founding mythologies.

As any anthropologist knows, that is not unusual. Almost every society in the world has some kind of cognitive map in place, shared cultural rituals and symbolic landscapes, which reinforce the position of the elites. And that cognitive map is often riddled with contradictions which go largely unrecognised, largely because [. . .] there are social silences. What people don't discuss is as important as what they do discuss, whether in financial circles or a Tadjik village.

So in a funny kind of way, my 'Anthropology 101' was absolutely critical in helping me understand not just what was happening inside finance in 2005 and 2006, but also in spotting some of the dangers and the limitations in the way bankers and financiers were looking at the world.

I often joke in retrospect that if only the banks had had a few anthropologists in their senior echelons, they might not have been quite so dumb. Not because anthropologists are necessarily any better at finance, but because anthropologists are less likely to 'drink the cool aid' than any other participants.[15]

So which gives greater foresight, depth or distance?

As an anthropologist and a financial journalist prepared to steep herself in arcane detail, arguably Gillian Tett had the benefit of both perspectives.

Dusko Doder's example suggests that when engaging in the difficult art of reading between the lines to disentangle hidden political developments, long experience and close observation win out. But when, rather than the death of yet another octogenarian Soviet leader, a moment of crisis is under way and previous experience is about to be overturned, a seasoned journalist needs to have the courage to throw out their past knowledge, and anticipate a far broader and more unexpected range of options.

Close scrutiny can also be a drawback. It can mean you are too close to an event, and blind to the bigger picture. Journalists holed up in the nearest desert town to the Algerian hostage crisis had the benefit of some local first-hand accounts and a few leaks from Algerian officials, but there were also many frustrations. On that occasion it felt as though it was easier to piece together the jigsaw puzzle from London.

Conclusion

In conclusion, although journalists may proffer foresight with reluctance, the subject generates interesting issues.

Sometimes a journalist may display extraordinary foresight in spotting a trend or shift when no one else does. It may be because they bring a fresh perspective which, as Gillian Tett puts it, means they can see flaws in others' assumptions because 'they haven't drunk the cool aid'. It may be because of a hunch, built on years of experience, a wide range of contacts and a final flash of insight which allows the last piece in the jigsaw to slot in. It may be because they can venture where others cannot, taking the risk to slip into Aleppo, for instance, to see for themselves how the civil war in Syria is unfolding.

New tools like data journalism, or tapping into social networks may bring new insights; although rational analysis and information gleaned from afar may have limits when it comes to understanding the vagaries of human nature.

But there are ethical issues to consider. When is it appropriate for a journalist to make a prediction which may change the course of events? How wary should they be of revealing information which could put themselves and others in danger?

And instead of an either/or outcome, should journalists perhaps seek to weigh probabilities, highlighting the difference between a certain and a likely prediction? In this way they would encourage their listeners, viewers, and readers to view the world not as a set of discrete separate events, but – whether it is a presidential election, an uprising, or a kidnap drama – a series of evolving processes, dependent on a stream of inter-locking circumstances.

As a rule, though, perhaps it is best not to expect too much of journalists when it comes to crystal-ball gazing. Their ambition is more modest: a hope that their reports and insights will stand the test of time and merit that description of their trade as 'the first rough draft of history'.

Further Reading

1. BBC News online, 'William Rees-Mogg, Former Times Editor, Dies', 29 December 2012, www.bbc.co.uk/news/uk-20864974
2. The famous lunchtime weather forecast for the BBC, 15 October 1987. Available on Michael Fish's own website www.michael-fish.com/
3. Royal Television Society Huw Wheldon Lecture 2012, Lyse Doucet, 'Can TV Journalism Survive the Social Media Revolution?'
4. From 'Lyse Doucet on the Arab Spring' released by the BBC College of Journalism, bbccojovideo, http://wapspot.mobi/tube/search/lyse-doucet-on-the-arab-spring?page=11
5. BBC Interview with Michael Fish, 15 October 2012, www.bbc.co.uk/news/uk-19923565
6. Andrea Canning's report, 9 November 2012, NBC Today programme.
7. The video and transcripts of both Web interviews with President Putin are available at: http://news.bbc.co.uk/1/hi/world/europe/1198631.stm http://news.bbc.co.uk/1/hi/talking_point/5153854.stm
8. Doder, Dusko, *Shadows and Whispers. Power Politics Inside the Kremlin from Brezhnev to Gorbachev* (Harrap: London) 1986, pp 5–23.
9. See *New York Times* article on the settlement of the libel case, 2 August 1996, available at: http://www.nytimes.com/1996/08/02/world/time-settles-libel-case-brought-by-a-reporter.html
10. Robert Peston's scoop on Northern Rock, BBC News Online, 13 September 2007, http://news.bbc.co.uk/1/hi/6994099.stm
11. Evidence given by BBC Business Editor Robert Peston to the House Of Commons Treasury Select Committee, 4 February 2009, http://news.bbc.co.uk/1/hi/uk_politics/7870240.stm
12. Michael Buerk's report on the Ethiopian famine, 23 October 1984, http://news.bbc.co.uk/1/hi/in_depth/8315248.stm

13. Ben Brown's report from the Dadaab refugee camp in Kenya, 4 July 2011, http://www.bbc.com/news/world-africa-14023160
14. See the so-called 'Charles Duelfer Report', the Comprehensive Report of the Special Advisor to the Director of Central Intelligence on Iraq's WMD, released September 2004, with Addenda released in March 2005, www.cia.gov/library/reports/general-reports-1/iraq_wmd_2004/index.html
15. Anthropology in the World Conference 2012 Keynote Lecture, Dr Gillian Tett, available at: http://youtu.be/QJKLqWiIh8k

3 Foresight and Fiction

ROBERT J. SAWYER

When we speak of 'foresight', we're using the trendy new term for what used to be called 'futurism'. But neither term is particularly explanatory. Yes, they both connote seeing ahead, but they don't describe how that is done. In my field of science fiction, we refer to what we do as 'extrapolation', and I think that's a much better term than either 'foresight' or 'futurism'. To extrapolate, according to the dictionary, is to 'infer or estimate by extending or projecting known information'. Science fiction is the only area of fiction in which this is routinely done, and of all of literature – fiction and nonfiction – it's the only field in which it's regularly done on long timescales and with such rigour.

Defining Science Fiction

Let's begin by defining science fiction, since it's a name often misused and poorly understood by those unfamiliar with the field. The term was coined by Hugo Gernsback, a Luxembourger who emigrated to the United States and founded the first science fiction magazine *Amazing Stories* in 1926. (Actually, his first stab at naming this field was the portmanteau word 'scientifiction', but that didn't catch on.) In any event, Gernsback's definition of science fiction was 'fiction about science'. Please note that it never was, and is not today, about *fictional* science. We do not just make stuff up. When a benighted newscaster or columnist says that something improbable is 'just science fiction' what he or she really means is that it's 'just fantasy'. Science fiction is about things that plausibly might happen; fantasy is about things that could never happen – magic and the supernatural have no basis in reality.

I will refer repeatedly to Arthur C. Clarke, who is my favourite science-fiction writer, even if he did make a glaring mistake. In 1965, he coined what's come to be known as Clarke's Third Law, and it goes like this: 'Any sufficiently advanced technology is indistinguishable from magic.' That's flat-out wrong: magic involves the violation of the laws of physics, most often the law of conservation of mass and energy; there are constraints on what even the most-advanced technology can do, and good science fiction acknowledges and works within those constraints.

It really is a shame that science fiction is so often shelved in the same section of the bookstore as fantasy; they are antithetical genres, but they are paired due to a historical oddity. The first US printings of J.R.R. Tolkien's *The Lord of the Rings* were pirate editions published by Ace, a science-fiction publisher. If science fiction had to share shelf space with another genre, it really should have been mystery fiction, as both prize picking up clues, deductive reasoning, and rational thought.

Anyway, Gernsback's definition, 'fiction about science', served well enough for a couple of decades. Then Isaac Asimov, the great Russian-born American science-fiction writer, broadened Gernsback's definition to this: 'Science fiction is that branch of literature that deals with the responses of human beings to changes in science and technology.' This revised definition put the science-fiction genre squarely in the foresight arena. I have coined two definitions of the genre myself. One is simply that 'science fiction is the literature of intriguing juxtapositions'. That is, it is the field in which you can find quantum computing and paleoanthropology cheek-by-jowl, as in my own novel *Hominids*, or where information theory, Chinese politics, primate communication, and the story of Helen Keller can all spark off each other, as in my novel *Wake*.

More germane to a discussion of foresight is my other definition: 'Science fiction is the mainstream literature of a plausible alternative reality.' 'Mainstream' in the sense that it's told as if to someone already familiar with the milieu of the story; if the story is set on Mars in the year 2087, the writer spins the tale as if the reader lives in 2087 and if he or she doesn't happen to be a denizen of Mars, it's no more exotic a locale to him or her than, say, Madagascar is to us.

Another of my criteria – an 'alternative reality' – is designed to include not just the future but alternate presents and pasts, the field of counterfactual history, exploring scenarios such as the Axis triumphing in World War II, as exemplified by Philip K. Dick's 1962 masterpiece *The Man in the High Castle*. American author Harry Turtledove is the modern master of alternate history. Such stories are widely regarded as a subset of science fiction, and for good reason: the same extrapolative skill is required in writing them; the only difference is that the point being extrapolated forward from is in the past rather than the present.

In Gernsback's definition, in Asimov's, and in my own one about the mainstream literature of alternative realities, the constant, whether explicitly stated or merely implicit, is plausibility. *Science* is the empirical, the verifiable, the falsifiable; *science fiction* accepts the scientific method as the only legitimate way of knowing. You won't find a more skeptical bunch about mental powers, past lives, UFOs, or New Age nonsense than science-fiction writers. A flat-out rejection of the supernatural or the implausible is only one of the fundamental building blocks for science fiction. The other foundational block is the core truth that science fiction espouses. Mainstream fiction – also known as *mimetic* fiction, since it is imitative of real life – takes as a given that you can't change human nature; science fiction takes as a given that you can. Yes, there are mainstream works that argue for attitudinal shifts – *Uncle Tom's Cabin* is a classic example – but such shifts are predicated upon appealing to existing, immutable human physiology and psychology. In Harriet Beecher Stowe's abolitionist novel, the appeal was to existing human compassion. It wasn't based on the assumption that a new capacity, namely one for empathy, could be developed in human beings when none had existed before. Charles Darwin understood that humanity is a work in progress.

Standing on Darwin's Shoulders: Humanity Changing Itself

Science-fiction writers have stood on Darwin's shoulders going all the way back to the days of H.G. Wells. The example from Wells's oeuvre of physical change that first springs to mind is from his 1895 novella *The Time Machine*, in which, 800,000 years from now, humanity has bifurcated into two species, the brutal, clever Morlocks and the feckless, feeble-minded Eloi. But much more interesting in this regard is my favourite Wells novel, *The Island of Dr. Moreau*, published the following year, in 1896, because in it change isn't something that happened to humanity; rather, change is something done with deliberation by humans. Dr. Moreau creates chimera beings, infusing animals with human traits – or perhaps vice-versa – producing creatures that think and feel fundamentally differently from us. And, of course, Aldous Huxley gave us one of our first tastes of genetic engineering, creating new kinds of humanity with his various castes grown in glass containers in his 1932 novel *Brave New World*. And, building on that, science fiction has long posited that fundamental human attitudes and mental capacities can and will change.

You can easily start an argument in any pub about which version of *Star Trek* is the best, but for me it will always be the original series with Kirk, Spock, and McCoy. That said, it spectacularly fails with regards to science fiction taking a position that you can change human nature. Although set in the twenty-third century, the crew of Kirk's *Enterprise* consisted of mid-twentieth-century people with twentieth-century mindsets. At one point in the episode *The Squire of Gothos*, Captain Kirk actually teases Spock by asking him if he'd ever 'dipped little girls' curls in inkwells'. Kirk and the rest were very much mired in their 1960s roots. Indeed, much of their behaviour doesn't even pass muster now. Certainly Kirk's womanizing would not be tolerated; you have to be at least a general to get away with that sort of thing today. And Dr. McCoy says things to Spock that we simply wouldn't countenance anyone saying in the workplace about a member of a different ethnic group: 'You pointed-eared, green-blooded hobgoblin!' We also see lots of alcohol abuse, and despite some noble efforts, a lot of sexism, too, including the statement that women couldn't command starships.

To his credit, all of this started to grate on *Star Trek* creator Gene Roddenberry. When it came time for him to write the novelization of *Star Trek: The Motion Picture* in 1979, he proposed that much of the human race had evolved into more advanced 'New Humans', and that Kirk and company were throwbacks, whose primitive impulses made them suited for gung-ho exploration. Eight years later, when Roddenberry introduced us to Jean-Luc Picard and Commander Data in *Star Trek: The Next Generation*, he decided to directly address the issue of changing human nature. He decreed that come the twenty-fourth century, interpersonal conflict would be a thing of the past – as would the excesses of alcohol. His new crew drank 'synthehol', the pleasing effects of which could be shrugged off with an effort of will, shocking the twenty-third-century Scotty who shows up in one episode.

The problem with having no interpersonal conflict was that, at least at the beginning, *Star Trek: The Next Generation* was boring, not to mention pretty much unwatchable. It was only after Roddenberry passed away that this constraint – something the show's writers had called 'The Box' – was done away with. The more science fiction tries to portray the future of humanity as a changed species, the harder it is for an untrained, present-day audience to identify with it. As critic Samuel R. Delany has rightly pointed out, reading serious science fiction is hard work, and with science fiction currently predicting everything from profound body modification (people with gills or extra arms or bigger brains or sonar or what-have-you) to radical life extension providing practical immortality; the boosting of mental powers and the addition of new senses; scanning consciousness and uploading it into a virtual computer world or downloading it into durable android bodies; and even the fusing of individual consciousness into hive minds, the task of making modern science fiction accessible to the general reading public is daunting. For instance, here's an early paragraph from Scottish writer Charles Stross's 2006 novel *Glasshouse*, published to considerable acclaim by Orbit in the United Kingdom and Ace in the United States:

> The Invisible Republic is one of the legacy polities that emerged from the splinters of the Republic of Is, in the wake of the series of censorship wars that raged five to ten gigaseconds ago. During the wars, the internetwork of longjump T-gates that wove the subnets of the hyperpower together was shattered, leaving behind sparsely connected nets, their borders filtered through firewalled assembler gates guarded by ferocious mercenaries. Incomers were subjected to forced disassembly and scanned for subversive attributes before being rebuilt and allowed across the frontiers. Battles raged across the airless cryogenic wastes that housed the longjump nodes carrying traffic between warring polities, while the redactive worms released by the Censor factions lurked in the firmware of every A-gate they could contaminate, their viral payload mercilessly deleting all knowledge of the underlying cause of the conflict from fleeing refugees as they passed through the gates.

Whew! My point is that the most elaborate extrapolations are, by their very nature, the least accessible texts. This is to be expected, of course. One of the most influential papers in modern philosophy is Thomas Nagel's 1974 *What Is It Like to Be a Bat?* in which he argued that it's impossible for beings like us, without sonar and without the ability to fly, to understand at all what those things would be like for beings that possess them. Another paper, 'Helen Keller as Cognitive Scientist', published in 1996 by Justin Leiber, likewise argued that it is impossible for us, as sighted, linguistic beings to comprehend the thoughts of young Helen Keller in her blind, pre-linguistic state. That paper struck me as a challenge, and led directly to my WWW trilogy about the World Wide Web gaining consciousness.

Still, no matter how hard it is to portray, this notion that human beings – and their social structures – are not static but rather can and will change radically is central to science fiction.

The Missing Americans: Socialized Medicine and Foreseen Societies

Let me go off on what seems like a digression for a moment, but I promise it's germane. I'm a Canadian, and Canadians for the last decade and a half have been disproportionately represented on the ballots for science-fiction awards, often making up 40 percent or more of the shortlist for the major ones. Indeed, the field's top award, the best-novel Hugo, went to Canadians three times in the past decade, astonishing when you consider Canada's small population of 35 million. In that same time period, the award also three times went to Brits, but Brits are much more numerous than Canadians. Still, that means the majority of recent best-novel Hugos have gone to writers from countries with socialized health care, and I don't think that's a coincidence; rather, the ability to become a full-time writer early on, without needing a regular job to provide medical insurance for oneself and one's family, allows writers in all fields to hone their talents in their twenties, an age at which their American colleagues are hoping someday to find the time to write.

Damon Knight, the founder of the Science Fiction and Fantasy Writers of America, once observed that the most unrealistic thing about science fiction was the preponderance of Americans in its stories; practically no one, he observed, is an American – and he's right: well over 90 percent of characters in science-fiction books are Americans, whereas less than 4 percent of humans really are from the United States. And Canada has a population just a tenth that of the United States. Why are Canadians, in particular, doing so well on a *per capita* basis with science fiction? I think one reason is that, as a nation, Canada does embrace that simple reality I spoke of earlier: you *can* change human nature, including fundamentally shifting attitudes or inculcating new capacities. Canadian essayist (and president of PEN International) John Ralston Saul has observed that one of the biggest differences between Canada and the United States is in Canada's approach to its foundational documents. Americans view their Constitution and Bill of Rights as holy writ, and expend an enormous amount of effort trying to make sure that twenty-first-century America lives up to – or down to – the ideals of a group of men mostly born in the early 1700s.

Canada, on the other hand, views its Constitution and Charter of Rights and Freedoms as works in progress: documents to be tweaked, changed, and, if ever the need should arise, completely rewritten, as humanity itself changes. That makes my compatriots particularly suited for the job of foresight through fiction – humanity *is* changing, and we Canadians acknowledge that. Canada's seventh prime minister, Sir Wilfrid Laurier, said in 1904, 'The twentieth century belongs to Canada.' I like to quip that he was off by a hundred years. I also think it's no surprise that the best nonfiction book documenting the fundamental change in human attitudes over time was written by a Canadian. I speak of *The Better Angels of Our Nature: Why Violence Has Declined* by Stephen Pinker, published in 2011.

Interesting fact: the Pentagon has double the number of washrooms it actually needs. Why? Because it was built in the 1940s, when the United States required separate 'Whites' and 'Coloreds' washrooms in public facilities. When I was born, in 1960, the United States was still a segregated country, with African Americans a downtrodden underclass. Before I'd turned fifty, a black man was sitting in the Oval Office. Although in academia, we often speak of the 'retire or expire' factor – the notion that a new idea, such as continental drift, can't become mainstream until the old guard is replaced by a new generation – in the United States, many of the same people who supported segregation came to recognize that it was wrong; *they* changed – as individuals – and so, collectively, society changed.

Extrapolating What Plausibly Might Happen

Still, whether you're an American, a Brit, or a Canadian – the three nationalities that produce most of the world's science fiction, and not just counting the work in English – how does one, in fiction, extrapolate to what plausibly might happen? The first thing you need, as I've stressed above, is the conviction that human nature *does* change, that our psyches and our societies are malleable. Indeed, it is this ability to change that may explain why we're here and all other forms of humanity have died out. Despite having bigger brains than us, Neanderthals were intellectually stagnant, making essentially the same stone tools – the Mousterian industry – for 200,000 years or so. Our kind of humanity, however, was constantly improving its technology – because our way of looking at the world was constantly changing. Linnaeus takes a lot of flak for hubristically naming our species *Homo sapiens* – people of wisdom – but if wisdom is the result of cumulative changing perceptions and perspectives then perhaps our nimble kind does deserve that name.

After accepting that human nature does change, the second thing you need to extrapolate the future is, perhaps ironically, a keen appreciation of history. I had the pleasure of interviewing the American science-fiction writer Kim Stanley Robinson in 1989 for a documentary series I was writing for CBC Radio. He made the point then that the only way to extrapolate a trend is to look not just at the present, but the past as well – the future is just the continuation of history. I agree with Robinson, and would add that it's the vector from past to present that gives directionality to our extrapolations. The plausible future is the one that continues past trends; implausible futures break off in new directions without sufficient cause. In fact, the standard story-generating template for science fiction is not, as many contend, simply asking 'What if?' – that is, merely having a neat idea and working out its consequences – but rather wondering what will happen if this, whatever this happens to be, goes on, projecting a trend to a logical extreme or natural end point.

The third requirement for effective extrapolation is a recognition that the rate of change is no longer linear, but rather is exponential. This is a notion that's been widely popularized by inventor Ray Kurzweil, including in his massive tome *The Singularity Is Near*, but the idea actually originated with a science-fiction writer, Vernor Vinge. Like many other science-fiction writers, Vinge was, at the time he wrote about this topic, also an academic; he has since retired from his position as professor of mathematics at San Diego State University, and he first put forward his ideas in an article rather than a story. As a well-regarded science-fiction writer, he was clearly using his science-fiction chops when he published his seminal essay 'The Coming Technological Singularity: How to Survive in the Post-Human Era', which came out in 1993, two years after Vinge took home the best-novel Hugo (the first of five Hugos he would eventually win) for his novel *A Fire Upon the Deep*.

And, yes, I do think it's fair for science fiction to take credit for the singularity notion, even if it wasn't published as science fiction, since it was first formulated in depth by a science-fiction writer. Likewise, I think science fiction can also claim geostationary communication satellites. They were first proposed by science-fiction writer Arthur C. Clarke, even though he chose to unveil his calculation that anything in orbit 23,000 miles above the equator would stay stationary in the sky, and that three such satellites could cover the entire surface of the Earth, in 1945 in the journal *Wireless World* rather than in a science-fiction story. In later years, by the way, Sir Arthur had a T-shirt that said, 'I invented the communications satellite and all I got was this lousy T-shirt.'

Anyway, the notion of the singularity is wrapped up in the idea that the rate of technological progress is accelerating. The classic example, of course, is Moore's Law, coined in 1965 by Gordon Moore of Intel, and usually formulated these days to say that computing power doubles every eighteen months. That means the computers we had ten years ago were only 1/128th as powerful as the ones we have today, and the ones we'll have ten years from now will be 128 times as powerful as our current best machines. Vernor Vinge told us the singularity – the moment when machine intelligence will exceed human intelligence – would arrive no sooner than 2005 and no later than 2030, a prediction, although twenty years old now, that still seems reasonable. And, at the moment it does arrive, a gigantic *woosh!* will occur, since thinking machines will be able to quickly engineer better thinking machines; very rapidly – perhaps in a matter of days or hours – humanity will be left far behind. Or, at least, that's what the singularitarians would have us believe.

Regardless of whether they're right or not, the key point (that the rate of technological advancement is accelerating) is one that must be grasped by any futurist. The amount of progress this decade will be much greater than in the last decade; the progress made this century will far outstrip that of the last century. And this *is* the century in which the human race will either go extinct or establish its stability for not just centuries but millennia to come. AIDS and cancer are tractable scientific problems. We lament our slow progress in conquering them, but we've only known the structure of DNA for fifty years now, and we've only had a map of the human genome for ten years. Also, we finally have computers powerful enough to deal with complex things such as protein folding. In other words, we finally have the tools, after 40,000 years of civilization, to do real medicine; we just got them, but the progress will be rapid. I'll be astonished if, by the hundredth anniversary of Crick and Watson's discovery of the structure of DNA here at Cambridge, that any diseases continue to be a serious threat to humanity.

And it's not just technological devices that are changing at an accelerating rate; rather, it is humanity itself. The changes in the last fifty years – the collapse of the Soviet Union, the general decline in violence, the fact that a smaller percentage of the human race is in armed conflict than ever before, the acceptance in Europe and the growing acceptance in North America that atheists can have a role in public life, the legal recognition in many jurisdictions of same-sex marriages, the growth of women's rights, the recognition of the injustices that have been done to aboriginal peoples, the end of colonialism, the end of segregation in the United States and South Africa, and the fact that the United Kingdom and Canada have had female prime ministers and that the United States and South Africa have had black presidents – all attest to the rapidity of societal change. And even more change will come in the next fifty years than what we experienced in the previous half-century.

Robert J. Sawyer

To see some failings in this area, let's turn to the poor stepchild of written science fiction: science-fiction film and television, what we in the business call 'media sci-fi'. Such fare represents a different realm, with different roots, and most of the good things that can be said about science fiction as an extrapolative genre apply only to written science fiction. One of my favourite examples of the failing of science-fiction films to recognize the increasing rate of technological and societal change is at the beginning of what is otherwise regarded as one of the best science-fiction films of the twentieth century, *Forbidden Planet*, which starred Canadian actors Leslie Nielsen and Walter Pidgeon. *Forbidden Planet* was released in 1956, one year before the launch of the first Sputnik, five years before Yuri Gagarin became the first man to orbit the Earth, and just thirteen years before we landed on the moon. But it begins with this notice: 'In the final decade of the 21st Century, men and women in rocket-ships landed on the Moon.' Sitting in the mid-1950s, and looking at how long it had taken us to get from steam engines to a car in every driveway (150 years) it seemed likely that the moon was that far in our future, if you assumed a steady rate of technological change.

The Future Doesn't Happen One at a Time: Kubrick's 2001

Our first three requirements for extrapolating were: (1) a recognition that human nature, and human societies, do change; (2) an appreciation of history – of what has gone before; and (3) an understanding that the rate of change is accelerating. The fourth thing a good extrapolator must remember is the dictum from the greatest American science-fiction editor, John W. Campbell, Jr.: the future doesn't happen one at a time. A case-in-point is what I consider to be the finest science-fiction film of all time: *2001: A Space Odyssey*, made in the United Kingdom, and released in 1968. Its screenplay was written by Arthur C. Clarke and Stanley Kubrick who also directed the film. Now, I'm the first to admit that *2001* made a few enormous blunders in its attempts to extrapolate technology three and a half decades ahead, and I'll come to the reason for that failing later. But its greatest failing was perhaps in the area of societal extrapolation, that is, Clarke and Kubrick thought *only* technology would change, and so *2001* depicted an

68

all-white future. Not a single person of colour appears in the movie. Indeed, so little thought was given to the international nature of the future that one of the first things you see after the first line of dialogue is spoken, thirty-seven minutes into the film, is a futuristic immigration computer that asks you to choose your native language from a list presented *in English* – offering English, German, French, Spanish, and so on, instead of English, Deutsch, Français, or Español.

The second major mistake was failing to acknowledge the increasing prominence of women. By the time the film was made, there had already been a woman in space, the cosmonaut Valentina Tereshkova. Not only that, but when *2001: A Space Odyssey* came out in 1968, *Star Trek*, with its multiracial crew including Lt. Uhura and many other females, had been on the air for two years.

The third major mistake in *2001* was thinking the Cold War would endure into the twenty-first century.

And the fourth major mistake, and this was one that no science-fiction writer got right, was the belief that humanity would continue its ever-outward expansion into space. By the time of *2001*, we were to have giant-wheeled orbiting space stations, a city on the moon, and crewed interplanetary missions. Instead, just three and half years after the first person walked on the moon the *last* person to do so did. In the year-end summations that came out for 2012, every newspaper and news magazine noted the passing of that first man, Neil Armstrong, but none remarked on the fact that the thirtieth anniversary of the last man on the moon also passed last year.

One can almost – almost – have sympathy with moon-landing deniers, who marvel at the notion that we could have done in the 1960s something we can't do in the twenty-teens. I was lucky enough to have dinner with Buzz Aldrin, the second man on the moon, a couple of years ago. He'd been lobbying for commercial airlines to let astronauts into their airport lounges for free. He'd been getting some pushback: the airlines were saying there were just too many astronauts these days. Buzz countered that the perk should then be limited to 'real' astronauts, the ones who had undergone trans-lunar injection (TLI; leaving Earth orbit to voyage to another world) of which there were precisely twenty-seven, the crews of *Apollos* 8 and 10 through 17. No one has undergone TLI – indeed, no one has gone more than 500 kilometres from Earth – since 1972, just four years after *2001: A Space Odyssey* debuted.

The mistake Arthur C. Clarke and Stanley Kubrick had made was the mistake every science-fiction writer had made: they'd assumed the whole human race shared their agenda, an agenda that basically said price was no object. But, of course, price was, and is. Despite my earlier comments about the prevalence of Canadians on the science-fiction award ballots, I contend that 1000 years from now, if you look up 'science fiction' in the *Encyclopedia Galactica*, the entry will begin, 'A 20th-century American literary genre'. No one in the nineteenth century dreamed of such profligate spending as depicted in the movie *2001*, nor can anyone in the twenty-first century. Indeed, although Thomas Carlyle dubbed economics 'the dismal science' in 1849, it was one area of extrapolation mostly ignored by science-fiction writers for a very long time. (Today, though, I'm pleased to announce that it's at the core of many fine works of extrapolation, including the *Unincorporated* series by brothers Dani and Eytan Kollin.)

Taking Extrapolation Too Far

Of course, it's possible to take extrapolation too far. As it happens, my father is professor emeritus of economics at the University of Toronto. And although it was he who took me to see *2001: A Space Odyssey* at a theatre during its first run in 1968, he had never read any science fiction. And yet as a bright gadget-loving scientifically and mathematically literate person, he perfectly fit the core demographic. So I set about to introduce him to the joys of the genre. My first couple of suggestions failed to strike his fancy, and then it hit me: the perfect choice. My father's specialty was economic forecasting; in fact, he had headed the University of Toronto's Institute for the Quantitative Analysis of Social and Economic Policy, which tried to gauge the effects government programs would have on the economy, and he had pioneered many techniques of econometric modeling, including, I vividly remember from my childhood, one that involved elaborate structures made out of Tinker Toys.

And so I gave my father Isaac Asimov's magnum opus, *The Foundation Trilogy* – the series that has confounded countless readers and librarians because the title of the third book in the series is *Second Foundation*. *The Foundation Trilogy* tells of Hari Seldon, a social scientist and mathematician who has developed a field he calls 'psychohistory' which predicts social trends over not just months or years, but millennia. Asimov's position was that when the human population gets big enough, no one person can have a significant impact, and so the laws of mass action will apply, letting the broad strokes be mapped out; by the time of his story there are quadrillions of human beings scattered over thousands of worlds. Unfortunately, Hari Seldon's psychohistorical analysis predicts that the vast galactic empire is about to fall, and a dark age of 30,000 years' duration will ensue. He sets out, more or less negating Asimov's premise that one person can't make a difference, to ensure that a new prosperous civilization with arise after the interregnum. The problem, of course, is that Asimov started writing *Foundation* around 1940, decades before the notion of chaos theory, and sensitive dependence on initial conditions, had been developed. We know now that even a slight change has gigantic effects; rather than being damped out, it can alter everything.

This is one case in which media science fiction actually got it more correct: the final episode of *Star Trek: The Next Generation* is called 'All Good Things. . .' and in it the omnipotent Q takes Captain Picard back four billion years to observe the primordial ooze on the young Earth. As he says, if he were to merely stir that ooze with his finger, humanity will never be born.

I took *2001: A Space Odyssey* to task earlier for some of its extrapolative failings – but, of course, it also had many resounding successes. Something very similar to the tablet computers we all use now was shown in this film made in 1968: a flat-screen device that could display any content. The name was even close to what the real product ended up being called: 'NewsPad' instead of 'iPad'. The only thing they got wrong is that in the film, you can clearly see that the manufacturer of this wonderful device was IBM. Apparently, Big Blue didn't have the foresight to actually try to make the devices portrayed in the movie, ceding one of the biggest technology booms of recent years to a rival. The most significant creation in *2001: A Space Odyssey*, though, was the Hal 9000 computer, voiced by the wonderful Canadian actor Douglas Rain.

Robert J. Sawyer

Extrapolating Computers

There's a category of book I call shop-floor sweepings: you look around for what's lying about, sweep it up, and collect it into a single volume. William Gibson's recent *Distrust That Particular Flavour* was one such, as was my own *Relativity: Essays and Stories.* Yet another was Arthur C. Clarke's *The Lost Worlds of 2001*, which collected fragments of earlier drafts of the novel written simultaneously with the screenplay. In this book, we learn that originally the spaceship crew was to be aided by an ambulatory robot named Athena; it was only quite late in the development process that the notion of a central computer came to the fore. The name Hal was said to be a contraction of Heuristically programmed ALgorithmic computer. However, it's also true that H-A-L is one step alphabetically ahead of I-B-M. Arthur C. Clarke denied that that was deliberate. I never used to believe him; the odds are hugely against it coming up by accident. But my own first novel, *Golden Fleece*, published in 1990, is very much a homage to *2001*, and it features a central computer named JASON, which someone pointed out to me could be rendered not only as J-A-S-O-N but also as J-C-N, and the letters J-C-N come one letter alphabetically after I-B-M.

In the movie *2001*, Hal says his birthday was 12 January 1992. Now, it's not actually clear what part of the movie takes place in the year 2001. After we leave the ape-men and go to the future, there's the portion of the film that takes place on the wheel-shaped space station and on the moon-base, and then there's the Jupiter mission that's identified as occurring eighteen months later. If we assume the Jupiter mission is the part set in 2001, then we're supposed to believe that a cutting-edge spaceship had an eleven-year old central computer – and if it's the moon-base stuff that's set in 2001, then Hal would be thirteen by the time the mission took place. Too late for the movie, Arthur C. Clarke realized this was a mistake; Moore's Law was coined in 1965, about when Clarke and Kubrick started collaborating on their screenplay, but Clarke apparently hadn't gotten the memo yet.

Still, for the novel, which was published after the movie was in theatres, he changed the date to 12 January 1997 (five years later) meaning Hal is between four and six years old when we meet him. To commemorate that date – 12 January 1997 – in real life, MIT Press published a lovely volume entitled *Hal's Legacy: 2001's Computer as Dream and Reality*. The book has contributions from or interviews with such seminal computing figures as Marvin Minsky, Ray Kurzweil, and Douglas Lenat, as well as cognitive scientist Daniel C. Dennett. The essays in the book make the case that the whole agenda of the computer-science and artificial-intelligence communities was set from 1968 for the next thirty years by the vision people saw in the movie *2001*.

And, indeed, it was. In *2001*, Hal beats a human at chess, exhibits speech recognition, exhibits natural language processing, and has very sophisticated vision, including the ability to read lips. He also shows a remarkable talent for facial recognition, not only recognizing real people but recognizing others in sketches done by an amateur artist, and he even seems to exhibit common sense and moral reasoning.

Arthur C. Clarke's famous peer was Isaac Asimov, whose name is indelibly associated with robots, thanks to his fictional Three Laws of Robotics. Those laws, as Asimov himself told me in a 1985 CBC interview, were actually coined by the great editor of *Astounding Stories* I mentioned previously, John W. Campbell, Jr., so it's no surprise to find that Asimov himself didn't really understand computers. Back in 1952, his attention was caught by UNIVAC, the computer that correctly predicted that Dwight Eisenhower would beat Adlai Stevenson in the race for the White House, when all the traditional pollsters predicted Stevenson would win. The name UNIVAC is a contraction of 'UNIVersal Automatic Computer', but Asimov figured it was a computer with one vacuum tube, and so decided that his futuristic fictional computer in his 1956 story 'the Last Question' would outdo it by having lots of vacuum tubes – leading him to dub his thinking machine Multivac. Multivac was never networked; it was a single giant physical entity, tended by hundreds of technicians, and if you wanted to ask it a question, you had to go to it.

Extrapolating the Internet

And, speaking of networks, it's a silly canard that Al Gore claimed to have invented the Internet. Of course, he never said that. It's an equally silly claim – although one often heard – that science fiction *failed* to predict the Internet. In fact, it did so repeatedly. The oldest reference to something like the Internet was probably Mark Twain's 'telectroscope', which he proposed in an 1898 short story: 'The improved "limitless-distance" telephone was presently introduced, and the daily doings of the globe made visible to everybody, and audibly discussable too, by witnesses separated by any number of leagues.'

Something even closer to our modern Internet, and the World Wide Web that supervenes upon it, was put forth in Murray Leinster's short story *A Logic Named Joe*, first published in 1946. When Murray Leinster published his story, the word 'computer' referred to a person who worked with a calculating machine; it was the name of the operator, rather than the device and so he needed a term for the actual machines, and he came up with Logics, which is pretty good. As for calling this particular one 'Joe', well really, is that any sillier a name than 'Google'? In any event, Leinster predicts massively interlinked computers providing answers to questions on any subject at any time from anywhere. His narrator, an aw-shucks repairman, describes technology eerily reminiscent of what we now rely on two-thirds of a century later:

> You know the logics setup. You got a logic in your house. It looks like a vision receiver used to, only it's got keys instead of dials and you punch the keys for what you wanna get . . . Say you punch 'Station SNAFU' on your logic. Relays in the tank take over an' whatever vision-program SNAFU is telecastin' comes on your logic's screen. Or you punch '*Sally Hancock's Phone*' an' the screen blinks an' sputters an' you're hooked up with the logic in her house an' if somebody answers you got a vision-phone connection. But besides that, if you punch for the weather forecast or who won today's race at Hialeah or who was mistress of the White House durin' Garfield's administration or what is PDQ and R sellin' for today, that comes on the screen too . . . everything you wanna know or see or hear, you punch for it an' you get it.

And let us go back to Arthur C. Clarke's novel version of *2001: A Space Odyssey*. In that book, he has one of his characters using the NewsPad I described earlier thus:

> When he tired of official reports and memoranda and minutes, he would plug in his foolscap-size newspad into the ship's information circuit and scan the latest reports from Earth. In a few milliseconds he could see the

headlines of any newspaper he pleased ... one could spend an entire lifetime doing nothing but absorbing the ever-changing flow of information.

In other words, science fiction even predicted that surfing the Web could become an enormous time sink!

Now, remember what I said about the *Foundation* trilogy earlier: Isaac Asimov had failed to take into account – because it hadn't been invented yet – the notion of chaos theory, which meant any prediction might go awry, thanks to ignored seemingly small effects. Although the most famous science-fictional prose treatment of computing is William Gibson's 1984 novel *Neuromancer*, is it any wonder that it doesn't accord with how reality turned out, given that Gibson wrote it on a manual typewriter?

I've sometimes said in interviews that my recent novel *Wake* (and its sequels *Watch* and *Wonder*) are in dialog with *Neuromancer*, but where Gibson's view is pessimistic and closed (a hacker underground and/or big corporations controlling everything) mine is optimistic and open (power devolves to all individuals everywhere). Gibson's take, fascinating when he first put it forth, has been superseded by reality; the whole cyberpunk fork of science fiction is now a kind of alternate history unrelated to how computing really evolved: instead of cyberpunks, we got the communal Wikipedia, and *Time* magazine naming 'You' – us, the average joe who freely and altruistically creates online content – its 2006 Person of the Year.

The difference between Gibson's approach and mine is driven home most directly in *Wake*, where I paraphrase the opening line of *Neuromancer*, then add a final clause that turns its meaning around. *Neuromancer* begins, 'The sky above the port was the colour of television, tuned to a dead channel.' When Gibson wrote 'the colour of television, tuned to a dead channel', he meant to imply a gray foreboding firmament – but technology changed in ways he didn't anticipate. In my novel, I write, 'The sky above the island was the colour of television, tuned to a dead channel – which is to say it was a bright, cheery blue.' *Neuromancer* is, of course, a remarkable achievement, but *Wake* came out twenty-five years later, and starts extrapolating forward from a reality in which the World Wide Web actually exists.

When Science Goes Wrong

Still, Gibson and I are in accord on some things. As I argued in a 1999 speech at the Library of Congress, the central message of science fiction is this: 'Look with a skeptical eye at new technologies.' Or, as Gibson has put it, 'the job of the science-fiction writer is to be profoundly ambivalent about changes in technology'. Now, certainly, there are science-fiction writers who use the genre for pure scientific boosterism: science can do no wrong; only the weak quail in the face of new knowledge. Jerry Pournelle, for instance, has rarely, if ever, looked at the downsides of progress. But most of us, I firmly believe, do take the Gibsonian view: we are not techie cheerleaders, we aren't flacks for big business or entrepreneurism, we don't trade in utopias. Neither, of course, are we Luddites. The late Michael Crichton used to write of the future, too, but he wasn't really a science-fiction writer; if anything, he was an anti-science writer. Indeed, both Gregory Benford and I discussed with our shared agent, Ralph Vicinanza, why it was that Crichton outsells us. Ralph explained that he could get deals at least approaching those Crichton gets if – and this was an unacceptable 'if' to both me and Greg – we were willing to promulgate the same fundamental message Crichton does, namely, that science always goes wrong.

Think about it: when Michael Crichton made robots, as he did in *Westworld*, they run amuck, and people die. When he cloned dinosaurs, as he did in *Jurassic Park*, they run amuck and people die. When he found extraterrestrial life, as he did in *The Andromeda Strain*, it runs amuck and people die. When he delved into nanotechnology, as he did in *Prey*, it runs amuck and people die. Crichton wasn't a prophet; rather, he pandered to the fear of technology so rampant in our society – a society, of course, which ironically would not exist without technology. His mantra was clearly the old B-movie one that 'there are some things man was not meant to know'.

The writers of real science fiction refuse to sink to fear-mongering, and, indeed, we have an essential societal role, one being fulfilled by no one else. Actual scientists are constrained in what they can say. Even those scientists lucky enough to have tenure, which supposedly ensures the right to pursue any line of inquiry, are in fact muzzled at the most fundamental economic level. They cannot speculate openly about the potential downsides of their work, because they rely on government grants or private-sector consulting contracts. The government is answerable to an often irrational public. If a scientist is dependent on government grants, those grants can easily disappear. And if he or she is employed in the private sector, well, then certainly Samsung doesn't want you to say cellular phones might cause brain cancer; Dow Chemical didn't want anyone to say that silicone implants might cause autoimmune problems; British American Tobacco didn't want anyone to say that nicotine might be addictive. Granted, not all of these potential dangers turned out to be real, but even considering them, putting them on the table for discussion, was not part of the corporate game plan; indeed, suppressing possible negatives is key to how all businesses, including those built on science and technology, work.

Extrapolating without Fear or Favour

There are moments, increasingly frequent moments, during which the media reports that 'science fiction has become science fact'. Certainly one of the most dramatic recent ones was made public in February 1997. Ian Wilmut at Roslin Institute in Edinburgh had succeeded in taking an adult mammalian cell and producing an exact genetic duplicate: the cloning of the sheep named Dolly. Dr. Wilmut was interviewed all over the world, and, of course, every reporter asked him about the significance of his work, the ramifications, the effects it would have on family life. And his response was doggedly the same, time and again: cloning, he said, had narrow applications in the field of animal husbandry.

That was all he *could* say. He couldn't answer the question directly. He couldn't tell reporters that it was now technically possible for a man who was thirty-five years old, who had been drinking too much, and smoking, and never exercising, a man who had been warned by his doctor that his heart and lungs and liver would all give out by the time he was in his early fifties, to now order up an exact genetic duplicate of himself, a duplicate that by the time he needed all those replacement parts would be sixteen or seventeen years old, with pristine, youthful versions of the very organs that needed replacing, replacements that could be transplanted with zero chance of tissue rejection. Why, the man who needed these organs wouldn't even have to go to any particular expense – just have the clone of himself created, put the clone up for adoption – possibly even an illegal adoption, in which the adopting parents pay money for the child, a common enough, if unsavoury, practice, letting the man recover the costs of the cloning procedure. Then, let the adoptive parents raise the child with their money, and when it is time to harvest the organs, just track down the teenager, and kidnap him, and – well, you get the picture. Just another newspaper report of a missing kid.

Far-fetched? Not that I can see; indeed, there may be adopted children out there right now who, unbeknownst to them or their guardians, are clones of the wunderkinds of Silicon Valley or the lions of Wall Street. Still, the man who cloned Dolly couldn't speculate on this possibility, or any of the dozens of other scenarios that immediately come to mind. He couldn't speculate because if he did, he'd be putting his future funding at risk. His continued ability to do research depended directly on him keeping his mouth shut.

The same mindset was driven home for me when I was co-hosting a two-hour documentary called *Inventing the Future: 2000 Years of Discovery* for the Canadian version of The Discovery Channel. I went to Princeton University to interview Joe Tsien, who created the 'Doogie Mice' – mice that were born more intelligent than normal mice, and retained their smarts longer. While my producer and the camera operator fussed setting up the lighting, Dr. Tsien and I chatted animatedly about the ramifications of his research, and there was no doubt that he and his colleagues understood how far-reaching they would be. Indeed, by the door to Dr. Tsien's lab, not normally seen by the public, was a cartoon of a giant rodent labeled 'Doogie' sitting in front of a computer. In Doogie's right hand is his computer's pointing device – a little human figure labeled 'Joe': the super-smart mouse using its human creator as a computer mouse.

Finally, the camera operator was ready, and we started taping. 'So, Dr. Tsien,' I said, beginning the interview, 'how did you come to create these super-intelligent mice?' And Tsien made a 'cut' motion with his hand, and stepped forward, telling the camera operator to stop. 'I don't want to use the word "intelligent,"' he said. 'We can talk about the mice having better memories, but not about them being smarter. The public will be all over me if they think we're making animals more intelligent.'

'But you *are* making them more intelligent,' said my producer. Indeed, Tsien had used the word 'intelligent' repeatedly while we'd been chatting.

'Yes, yes,' he said. 'But I can't say that for public consumption.'

The muzzle was clearly on. We soldiered ahead with the interview, but never really got what we wanted. I'm not sure if Tsien was a science-fiction fan, and he had no idea that I was also a science-fiction writer, but many *science fiction* fans have wondered why Tsien didn't name his super-smart mice 'Algernons' after the experimental rodent in Daniel Keyes's story *Flowers for Algernon*, made into the movie *Charly*, starring Cliff Robertson. Tsien might have been aware of the reference, but chose the much more palatable 'Doogie' – a tip of the hat to the old TV show *Doogie Howser, M.D.* about a boy-genius who becomes a medical doctor while still a teenager – because, of course, in the story *Flowers for Algernon* the leap is made directly from the work on mice to the mind-expanding possibilities for humans, and Tsien was clearly trying to restrain, not encourage, such extrapolative leaps.

We science-fiction writers also aren't bound by nondisclosure agreements, the way so many commercial and government scientists are. Because of that, we were the first to weigh in on the dangers of nuclear power (as in Lester del Rey's 1942 story *Nerves*). And we began the public discourse about the actual effects of nuclear weapons (as in Judith Merril's 1948 story *That Only a Mother* which deals with gene damage caused by radiation). Science fiction is the WikiLeaks of science, getting word to the public about what cutting-edge research really means.

And we come with the credentials to do this work. Many science-fiction writers, such as Gregory Benford, are working scientists. Many others, such as Joe Haldeman, have advanced degrees in science. Others still, such as myself, have backgrounds in science and technology journalism. Our recent works have tackled such issues as the management of global climate change (as in Kim Stanley Robinson's *Forty Signs of Rain* and its sequels), biological terrorism (as in Paolo Bacigalupi's *The Windup Girl*), and the privacy of online information and China's attempts to control its citizens' access to the World Wide Web (as in my own *Wake* and its sequels). And although one can't imagine George Lucas being asked to advise the space program, print science-fiction writers often do consulting for government bodies. A group of *science fiction* writers called SIGMA frequently advises the US Department of Homeland Security about technology issues, and Stephen Baxter, Allen Steele, and I were recently consulted by DARPA, the US Defense Advanced Research Projects Agency, about future spaceship designs.

Why do they come to us? Because someone needs to openly do the speculation, to weigh the consequences, to consider the ramifications – someone who is immune to economic pressures. And that someone is the science-fiction writer.

Although, as I mentioned, Isaac Asimov is most famous for the Three Laws of Robotics, in 1974 he coined his Three Laws of Futurics, and they define well the science-fiction approach to extrapolation. The Laws of Futurics are:

1 What is happening will continue to happen.
2 Consider the obvious seriously, for few people will see it.
3 Consider the consequences.

However, we science-fiction writers don't just consider the obvious consequences: our job is not to see just the first-order effects, but the second- and third-order effects as well. Anyone could have predicted the automobile, but only a science-fiction writer would have predicted the traffic jam. Anyone could have predicted the airplane, but only a science-fiction writer would have predicted hijacking, frequent-flyer miles, and airport lounges.

This sort of extrapolation to second- and third-order effects goes right back to the very beginning of the genre. Brian Aldiss and many other critics contend that the first science-fiction work was Mary Shelley's 1818 novel *Frankenstein; or, The Modern Prometheus*. It explores, in scientific terms, the notion of synthetic life: Dr. Victor Frankenstein studies the chemical breakdown and putrefaction that occurs after death so he can reverse it to animate nonliving matter. Take out his scientific training, and his scientific research, and his scientific theory, and, for the first time in the history of fiction, there's no story left. Like so many other works of *science fiction* that followed, Shelley's story is a cautionary tale: it raises profound questions about who should have the right to create living things, and what responsibility the creators should have to their creations and to society.

Think about that: Mary Shelley put these questions on the table almost two centuries ago, forty-one years before Darwin published *The Origin of Species* and 135 years before Crick and Watson figured out the structure of DNA. Is it any wonder that Alvin Toffler, one of the first futurists, called reading science fiction the only preventive medicine for future shock? (Note *Frankenstein*'s publication date, by the way: 1818. The science-fiction bicentennial is just a few years from this lecture. I, for one, am going to have a party.)

When people talk about all the things science fiction correctly predicted, they often cite moon landings and submarines (suggested by Jules Verne) or surveillance technology (made famous by George Orwell) or the cell phone (inspired by *Star Trek*'s handheld communicator) or robots (the very name of which comes from a work of science fiction, Karel Capek's 1920 play, *R.U.R.*). That said, science fiction's job is not to predict the future. Rather, it's to propose and explore a smorgasbord of possible futures so that society can make informed decisions about where we want to go. George Orwell's science-fiction classic *Nineteen Eighty-Four* wasn't a failure because the future it predicted didn't turn out to be anything like the real year 1984; rather, it was a resounding success because it helped us avoid that fate. As Ray Bradbury famously said, 'My job isn't predicting the future; it's *preventing* the future.'

Still of all the things science fiction has foretold, I think the most important one is the simple fact that there will *be* a future. From the advent of nuclear weapons (the exact mechanism the secret Manhattan Project had in mind was predicted in such exquisite detail in the science-fiction magazine *Astounding Stories* that the FBI demanded a recall of one of its issues), through the Cold War and the war on terror, to the present day as we stand on the brink of catastrophic climate change, science fiction has always said – and continues to insist – that humankind *does* have a future, a future that stretches far ahead for hundreds, thousands, and even millions of years. This deeply held conviction that the human journey has only just begun is the most important, and the most wondrous, prediction of all.

4 Foresight in Scientific Method

HASOK CHANG

Introduction

My brief is to discuss foresight in relation to science. I don't think it is difficult to justify the inclusion of this topic in this year's series. Of all systems of practices and beliefs that humans have created over the centuries, modern science is widely regarded as the one with the best-ever claim to foresight. Scientists have used their theoretical insights to make impressive predictions of future events, which have contributed enormously to the prestige and authority of science. If foresight was equated with wisdom then it is no surprise that scientists gained their status as wise men.

Initial successes often involved predicting when certain striking phenomena would occur, demonstrating the understanding of natural patterns that govern phenomena that had seemed mysterious and supernatural before. The predictions of lunar and solar eclipses are salient examples, and there is also the famous case of Edmond Halley's prediction about the return of the comet named after him. More mundane yet highly significant successes also followed, such as improved weather forecasting and medical prognosis.

The foresight displayed by science also went well beyond pattern recognition. As the physical sciences matured, previously unforeseen kinds of entities were predicted by theory, and many such predictions were gloriously vindicated. The prediction of the existence of new planets, particularly Neptune, was a striking-enough success. Then came more bizarre and wonderful successes, such as the discovery of new chemical elements predicted by Dmitri Mendeleev's periodic table, the realization of electromagnetic waves predicted by James Clerk Maxwell's electrodynamics, and the detection of a whole series of new elementary particles predicted by theories of modern physics, the latest instance of which is the Higgs boson. Not only new entities, but unexpected behaviour of familiar entities, predicted under particular conditions specified by theories, made great impressions on people's minds. Einstein's prediction of the bending of starlight passing by the sun literally made news headlines around the world. Physicists became fearsome sages of the postwar world by predicting the possibility of the atomic bomb, and then delivering the actual thing according to plan.

I could go on and spell out the various kinds and instances of foresight displayed by science, and probably be able to make a relatively informative and amusing chapter out of that. But my aim is more critical. Accepting that modern science has displayed an impressive degree of foresight about nature, I want to ask what the precise significance of predictive success is within the context of scientific methods. That question will occupy roughly the first half of the chapter. After that I will raise a slightly different question about foresight relating to science: can we foretell the future of science itself? I will answer that question mostly in the negative, and close by discussing the implications of that answer. So here is the list of major questions that I propose to deal with:

1   Can science foretell the future?
2   What is the significance of predictive success?
3   Can we foretell the future of science?
4   What is the significance of the uncertainty of the future?

The first I have answered with a fairly simple 'yes', at least a qualified 'yes' meaning 'much better than any other system of knowledge or belief known to humanity'. I now move on to the second question.

The Significance of Foresight in Scientific Method

Predictionism

Many philosophers and scientists have especially valued the ability of science to make novel predictions, sometimes to the point of regarding it as the defining characteristic of science. This view, which I will call 'predictionism', has a long history, but in modern times it is most clearly revealed in the philosophy of Karl Popper and Imre Lakatos. In Popper's falsificationist methodology of science, only failed attempts to refute a theory can count as corroborating evidence for it; the more severe the test, the higher the evidential value. The basic predictionist intuition is that anyone can make up a theory to fit known data, but the true test of a theory is correctly predicting something unknown. The true scientific attitude according to Popper and other predictionists is the willingness to give up theories that make predictions that turn out to be false. And the predictions are no good unless they take real risks.

> Pseudoscientific theories attempt to accommodate all known observations by *ad hoc* manoeuvers, to avoid falsification at any cost. Scientific theories 'stick their necks out' by making bold predictions, which are then honestly checked against experience.
>
> (Popper, 1972)

Vague predictions that are compatible with anything that happens are a sure sign of pseudoscience, as in a newspaper astrology which will tell you that if you are Aries you 'may find love in an unexpected place', et cetera.

For the young Popper, spending his formative student years in the desperate ferment of Vienna in the aftermath of defeat in the First World War, battling pseudoscience was very serious business. He railed against the Marxists who had no trouble explaining every occurrence in society according to their grand theory of history, and then twisted their theory to accommodate happenings that went against their own predictions. This was exactly what they had done in the face of the first communist revolution which took place in agrarian Russia, not in a place that had reached the height of capitalist development as their theory predicted. He also attacked in the same vein the popular psychological theories arising in Vienna at the time: namely Sigmund Freud's psychoanalysis and Alfred Adler's 'individual psychology'. Popper actually worked for a time with Adler, assisting him in the social guidance clinics that Adler had established for working-class children and youth. He was at once impressed and disturbed by Adler's ability to interpret all psychological phenomena according to his own theory. He reports one particular personal experience: 'Once, in 1919, I reported to him [Adler] a case which to me did not seem particularly Adlerian, but which he found no difficulty in analyzing in terms of his theory of inferiority feelings, although he had not even seen the child. Slightly shocked, I asked him how he could be so sure. "Because of my thousandfold experience," he replied; whereupon I could not help saying: "And with this new case, I suppose, your experience has become thousand-and-one-fold." What I had in mind was that his previous observations may not have been much sounder than this new one; that each in its turn had been interpreted in the light of "previous experience," and at the same time counted as additional confirmation. What, I asked myself, did it confirm? No more than that a case could be interpreted in the light of a theory'.

The great contrast to all this was Albert Einstein's general theory of relativity. The year 1919 was important on this side, too, as that was when the British astronomer Arthur Eddington led the expedition to observe a solar eclipse, to test one of the striking predictions of Einstein's theory. Many readers will be already familiar with this story, but let me just explain the basic idea, for those of you who aren't (Figure 4.1).

The general theory of relativity stated that the path of light passing by a massive body would be bent. For example, starlight passing near the Sun would be bent on its way to the eyes of an earth-bound observer, creating a distortion in the apparent position of the star because our perceptive apparatus implicitly assumes that light travels in straight lines. This prediction can be checked by comparing the relative positions of stars right around the sun with the relative positions of the same stars observed at night without the sun near them, but it is impossible to see the stars in daylight to confirm their positions. A total eclipse of the sun provides a rare opportunity for seeing the stars around the sun. So Einstein's theory stuck its neck out, which was mercifully not chopped off. For Popper this was the best of science. Pseudoscience busies itself with making sense of already known facts. Real science is the business of foresight – exploring and expanding the frontiers of knowledge by fore-telling what we don't know yet. Modern science has demonstrated great foresight, and this is what makes it so great.

Popper's predictionism was elaborated and refined in instructive ways by the Hungarian philosopher Imre Lakatos. In agreement with Popper's basic point, Lakatos declared that the essence of scientific progress was

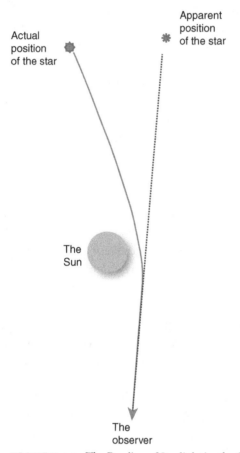

Actual position of the star

Apparent position of the star

The Sun

The observer

FIGURE 4.1   The Bending of Starlight by the Gravitational Field of the Sun

the making of successful novel predictions. But he also clarified that what may seem like an *ad hoc* move of accommodation can be a perfectly legitimate and progressive scientific move; in fact he thought that the protection of a favourite hypothesis was the standard mode of work in a scientific research programme (Figure 4.2). But Lakatos emphasized that modifications in the protective belt made in order to protect the theory against refutation must have their own testable consequences, constituting novel predictions. If there are no such novel predictions, then the modifications are *ad hoc* and to be denounced as non-progressive and unscientific.

Protective
belt

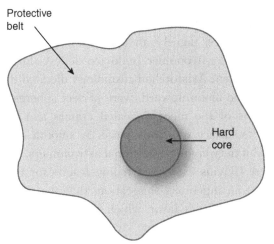

Hard
core

FIGURE 4.2   A Graphic Representation of a Lakatosian Research
Programme

A good example illustrating Lakatos's viewpoint is the discovery of the
planet Neptune, which arose from an obstinate adherence by astronomers
to Newtonian theory in the face of its apparent refutation. After the planet
Uranus was discovered in the late 18th century, astronomers noted that
its observed orbit was not in accordance with Newtonian gravitational
theory. Instead of taking this as a refutation of Newton's theory (and
thereby discarding what had been the best-ever theory in the entire
history of mathematical physics, arguably in all of science) astronomers
reasoned that Newton's theory *must* be right; therefore there was some
other assumption involved that was wrong. In the end they pinpointed
the implicit assumption that Uranus was the outermost planet in the solar
system, and hypothesized that there must be another, as yet unknown,
planet out there that was disturbing the orbit of Uranus with its own little
gravitational pull. Using Newtonian theory itself, astronomers calculated
what mass and orbit this hypothetical planet must have, and training their
telescopes at the predicted place, they found it, which they subsequently
named 'Neptune'. It may seem that Neptune was an *ad hoc* hypothesis,
invented solely for the purpose of protecting Newtonian theory from
refutation. And it was, but according to Lakatos it was a perfectly legitimate and productive move, because the Neptune hypothesis did lead to

novel predictions about what was observable in the heavens, which were happily confirmed. This can be contrasted to another, lesser-known case from the history of science, namely an opposition to Galileo's lunar observations by the Jesuit astronomer Ludovico delle Colombe. Up to Galileo's time, the prevalent Aristotelian cosmology dictated that all the celestial bodies, from the moon upward, were perfect spheres. Galileo's telescopic observations of the moon revealed craters and mountains, demonstrating that its surface was not a perfectly smooth sphere. This caused quite a bit of anxiety among traditional astronomers.

When Christopher Clavius canvassed fellow Jesuits for their views, Colombe replied with an ingenious suggestion, that the surface of the moon was covered in an invisible layer, which was perfectly smooth and spherical; Galileo was only seeing some strange inner structure of the moon. Colombe's hypothesis only protected the Aristotelian hard core, without at the same time generating any testable novel predictions short of someone going to the moon and checking for the invisible over-layer. As I mentioned, this kind of insight about the value of novel predictions was not new to Lakatos or even Popper. A very old predecessor can be found in René Descartes, who said that we know that our hypotheses are correct 'only when we see that we can explain in terms of them, not merely the effects we had originally in mind, but also all other phenomena of which we did not previously think'.

Logicism

After hearing Popper and Lakatos, with echoes of Descartes, one may become convinced of predictionism. But as it usually goes in philosophy, there is an opposing position that is equally convincing. This is a position that I call 'logicism', the basic intuition for which is that confirmation is a matter of logical relationship between hypothesis and evidence; the temporal order in which they are arrived at is irrelevant. Contemporary philosopher of science Laura Snyder expresses what she calls the 'objective concept of evidence' in the following way: 'whether e is evidence for h does not depend upon anyone's beliefs or knowledge about e, h, or anything else. Hence if some e is evidence for h, it is so regardless of what any person knows or believes.' So it should not matter whether scientists first formulated *h* and then found *e*, or vice versa. All that matters is that *h* 'predicts' *e* in the logical sense, meaning that we can deduce *e* from *h*. Her concluding words: 'Evidence is evidence is evidence.'

Like predictionism, logicism has a long and distinguished history. John Stuart Mill's comments might well have been directed at Popper almost a century later: 'Such [novel] predictions and their fulfillment are, indeed, *well calculated to impress the uninformed...* But it is strange that any considerable stress should be laid upon such a coincidence by persons of scientific attainments.' The exact same point is voiced by the Cambridge economist John Maynard Keynes: The 'peculiar value of [novel] prediction... is altogether imaginary... The question as to whether a particular hypothesis happens to be produced before or after examination of [its instances] is quite irrelevant.'

It is quite difficult to argue against the general point voiced by Snyder, Mill, and Keynes. There are also some specific circumstances gathered from the history of science that add to their scepticism regarding predictionism. For one thing, there are many perfectly respectable sciences which do not produce much by way of testable novel predictions (e.g. palaeontology, evolutionary biology, geology, cosmology). It is also difficult to deny that 'used' evidence has real evidential value; for example, I think it is obvious that Galileo's law of free fall and Kepler's laws of planetary motion, which Newton clearly knew about and used in making his theory of gravitation, still carry significant weight as positive evidence for Newton's theory.

Brush on the Value of Accommodation

This debate has the shape of a classic philosophical dilemma: both sides seem right, but they cannot be reconciled with each other. To make things worse, there is a surprising third view, from the historian Stephen Brush, who has argued that the accommodation of previously known facts can be *more* convincing evidence for a theory than the prediction of new facts. Brush says: 'a successful explanation of a fact that other theories have already failed to explain satisfactorily (for example, the Mercury perihelion) is more convincing than the prediction of a new fact, at least until competing theories have had a chance (and failed) to explain it'.

Brush's view, which I find quite persuasive, is that the really impressive thing about Einstein's general theory of relativity was not the prediction of light-bending (discussed above) but how it explained an old and well-known discrepancy in Newtonian gravitational theory, namely that the observed orbit of Mercury shifted around to a greater degree than predicted by Newtonian theory. There had been many attempts to resolve this anomaly, without conclusive success, until Einstein. Brush's perspective on this matter fits in quite well with Thomas Kuhn's view that a new paradigm gains credibility most importantly by resolving persistent anomalies of the previous paradigm.

The Verdict

So where does all of that leave us? When I teach this topic to my students, I leave the question unresolved, so that they can arrive at their own answers. Well, that is what I tell them, and it is true, but another part of the truth is that I've never had a definite answer to give them even if I wanted to, until now. In the course of preparing for the lecture I considered this issue all over again, and have arrived at an answer that is fairly satisfactory to myself. The first thing to say is that the confirmation of theory by empirical evidence is a complicated business, and there are many aspects of evidential support for a theory. The power to make novel predictions is just one of them. Here it is helpful to review the very sensible introductory overview that Carl Hempel gave of the empirical confirmation of scientific theories. He lists many 'criteria of confirmation and acceptability', and being confirmed by novel predictions occurs as just one of six such criteria.

1 Strength of evidence
   (a) Quantity
   (b) Variety
   (c) Precision
   (d) Novel predictions
2 Acceptability/credibility of hypothesis
   (a) Theoretical support
   (b) Simplicity

But I would take this one step further than Hempel, and say that the criterion of novelty is in fact only secondary, in the sense that its importance only comes from the fact that it can help the satisfaction of another criterion, namely the variety of evidence. I agree with the logicist argument that the novelty of prediction, in itself, has no special value in the confirmation of theories. But novelty is valuable because it ensures independence. Testing a theory through a novel prediction made by it means that the evidence gathered in the testing process is guaranteed to be different from the evidence that was originally used in inspiring the theory. This ensures, if the outcome is positive, that the variety of evidence supporting the theory will have increased. So that is good, but let's put the virtue of novelty in its place. Novel prediction is not the only way to achieve variety of evidence; it is merely a very striking way. For one thing, we may make use of facts that were already known but not considered in the making of the theory in question.

This point is actually already evident in the statement by Descartes, which I quoted earlier as an expression of predictionism. Descartes speaks of 'phenomena of which we did not previously think'. Now I think you will be able to see that this is more properly read as an advocacy for a variegated evidence base, and it says nothing about the temporal order in which the evidence and the hypothesis have come into being. Various modern philosophers, such as Alan Musgrave and John Worrall, have consolidated Descartes's insight into the doctrine of use-novelty (or the heuristic view of evidence). Without going into details of that discussion, let me just state my view that the point of use-novelty, too, is not novelty as such, but the achievement of evidential variety. This may sound odd, but foresight is not a virtue in itself; it is a virtue only when it helps us achieve other aims.

A Summing-up Example: Mendeleev's Periodic Table

This is often told as a classic tale of the triumph of a theory by successful novel predictions. Mendeleev's prediction of several new elements was indeed impressive. But was this handful of novel predictions more important than the impressive systematic accommodation of many dozens of known elements? And something we don't hear about so much is the fact that about the same number of other elements predicted by him (including 'ether') were never found! But this did not destroy the periodic enterprise. On the other hand, the eventual accommodation of the noble gases, which Mendeleev had not predicted and could not fit sensibly into his table for some time, also provided evidence for the periodic scheme that was as important as the novel predictions. The addition of the noble-gas column, which removed a very threatening anomaly, was actually a key moment in the acceptance of the periodic system.

Foresight about Science: Can We Foretell the Future of Science?

The 'End of Science'

At this point let me pull back to the overall plan of this chapter:

1   Can science foretell the future?
2   What is the significance of predictive success?
3   Can we foretell the future of science?
4   What is the significance of the uncertainty of the future?

So far I have addressed the first two questions that I posed, and concluded that modern science is indeed very successful at making novel predictions, but this acknowledged ability is only as valuable as its ability to organize and explain phenomena discovered in other ways. Let me now address the next question on the list, which concerns fore-sight *about* science, rather than foresight *in* science: can we foretell the future of science itself? And I am going to argue that scientists and others have been very unsuccessful in predicting the course of the development of science itself. Many eminent scientists have got this badly wrong. It has even been said that science-fiction writers are better than scientists themselves in predicting technological futures – and even their interest-ing predictive success has been limited to technology, not the theoretical aspects of science.

Perhaps the best-known and most embarrassing variety of foresight failure by scientists is various predictions of 'the end of science'. For example, various physicists of the late Victorian era are known to have stated that all the important things had been discovered in physics, and only the details were left to be worked out. The acute irony is that they made these statements just on the eve of the great upheavals brought by the advent of relativity and quantum mechanics. One of these unfortunate physicists was American experimentalist Albert Michelson, who would bring the first-ever Nobel Prize to the United States in 1907 for his work on precision experiments in optics. In 1894, at the dedication of the Ryerson Physical Lab at the University of Chicago, which he designed and directed, Michelson declared: 'The more important fundamental laws and facts of physical science have all been discovered, and these are now so firmly established that the possibility of their ever being supplanted in consequence of new discoveries is exceedingly remote ... Our future discoveries must be looked for in the sixth place of decimals.' Aside from the unfortunate timing of his remark, there is an additional layer of irony in Michelson's statement. He was the Michelson of the 'Michelson–Morley experiment', the best attempt up to that point to measure the speed of light in different directions, from which the velocity of the earth's motion with respect to the ether could be inferred. It was his own work that created a crisis of confidence in the reality of the ether and prepared the ground for the acceptance of Einstein's special theory of relativity, which dispensed altogether with the need for the ether.

A subtler, but equally interesting case is that of William Thomson, better known as Lord Kelvin, Cambridge-educated professor of natural philosophy at Glasgow. In 1900 Kelvin wrote: 'The beauty and clearness of the dynamical theory, which asserts heat and light to be modes of motion, is at present obscured by two clouds.' These 'clouds' were detailed difficulties in understanding the movement of bodies in ether and the spectral distribution in black-body radiation. Now, Kelvin thought of these clouds as remaining anomalies to be resolved before the system of classical physics could be completely consolidated. Instead, they turned out to be major instigators of the upheaval that overthrew classical physics completely. At least Kelvin did notice that these clouds were there and not going away easily, but who can tell for sure what clouds will move away or dissipate, and what clouds will bring us a thunderstorm? That is just the point.

Hasok Chang

The Pessimistic Induction

I do not cite such cases in order to denounce the particular physicists involved for failures of judgement. I would say that their judgements were in general as good as anyone else's. Rather, what the two cases illustrate is a general phenomenon: it is extremely difficult to predict the future course of scientific development. There is one specific feature which has been particularly worrying to philosophers of science, namely the fact that very successful theories are very often rejected by later scientists as false. Philosopher of science Larry Laudan has turned this observation into a famous sceptical doctrine, which we now refer to as the 'pessimistic induction (or meta-induction) from the history of science', which is one of the standard arguments presented against scientific realism. Here is a nice, concise formulation of the pessimistic induction, due to Stathis Psillos (1999):

> The history of science is full of theories which at different times and for long periods had been empirically successful, and yet were shown to be false in the deep-structure claims they made about the world. . . Therefore, by a simple (meta-)induction on scientific theories, our current successful theories are likely to be false. . . Therefore, the empirical success of a theory provides no warrant for the claim that the theory is approximately true.

The pessimistic induction implies that looking at what is successful in today's science will give us no reliable clue as to what will continue to be accepted in tomorrow's science.

Laudan gave a long list of examples ranging from the humeral theory of medicine to 19th-century ether theories. One might argue that many of the theories on his list were not all that successful. Against that argument, I have spent much of my life as a historian of science in demonstrating the cogency and power, even predictive success, of some of the long-discarded scientific theories such as the caloric theory of heat and the infamous phlogiston theory. But the point I want to make can be made with other examples that will be much more immediately palatable. Let's just take two of the acknowledged greats of Cambridge science, Maxwell and Newton. Newton's mechanics and Maxwell's electrodynamics were probably the two most successful scientific theories ever, before the 20th century. But they are now said to have been based on some fundamental

94

falsities, and they have been superseded by 20th-century theories such as relativity, quantum electrodynamics, and beyond. Can we not very easily imagine a not-so-distant future in which the superstring theories and the Big Bang Theory are rejected as once-successful yet false theories? It is often argued that the march of science is continuous and cumulative, even across apparent revolutions, because earlier theories are shown to be approximations to the later ones. But there is at least a remaining worry, voiced by Kuhn, that the fundamental ontology of the universe as told by our best science seems to be unstable, even when the numerical formulas develop thankfully in a convergent manner.

The Problem of Unconceived Alternatives

Looking forward rather than backward in time, we can note the same kind of problem of lack of foresight about the course of development of science. Kyle Stanford makes a persuasive historical–philosophical case for what he calls 'the problem of unconceived alternatives' for scientific realism, in his 2006 book *Exceeding Our Grasp*. Stanford notes that scientists are often defeated in their attempt to find the true theory by choosing the best one from all the different possible theories, because their imagination is understandably limited and they are not able to conceive of what is later shown to be the right story. He observes: 'Past scientists have routinely failed even to conceive of alternatives to their own theories and lines of theoretical investigation, alternatives that were both well-confirmed by the evidence available at the time and sufficiently serious as to be ultimately accepted by later scientific communities.' The philosophical point is simple and convincing: choosing the best available explanation gives no guarantee of the truth of the chosen explanation, unless we know that all possible explanations are available to us. That is precisely what the problem of unconceived alternatives prevents.

Stanford gives detailed examples drawn from the history of science, which I would recommend for everyone. For now, I will give you a very simple and brief illustration from recent science. Most of the scientists trying to ascertain the cause of bovine spongiform encephalopathy (BSE), or 'mad cow disease', were still tied to the idea that carriers of infectious diseases were all living organisms containing DNA or RNA, including bacteria, viruses, and others. But no one until recently conceived what is the more-or-less accepted theory now, namely that prions, misshapen proteins that can duplicate themselves, are responsible for BSE and other related diseases. So anyone trying to foretell the future of BSE-science would have got that badly wrong. We can say very similar things in relation to almost any other major developments in science. Would 19th-century physicists have been able to imagine non-Euclidean curved space-time? Would cosmologists fifty years ago have been able to imagine dark matter and dark energy, which are now supposed to make up a clear majority of substance in the universe? Without being able to conceive of such things, what predictions would they have made about the future of our knowledge of the most fundamental aspects of the universe?

What Is the Future of Science?

So, what can we say about the future of science? One short answer is that we don't know what will remain stable and what will be overturned. One important feature of that answer is that certainty, especially fanatical certainty, on the part of the scientists who believe a given theory does not seem to correlate with its longevity, not to mention its truth. And it seems the most fundamental parts of theoretical science that seem the least stable are stable in the long run. Some phenomenological laws have stayed good for many centuries, such as Snell's law of refraction or Archimedes's law of the lever. It is difficult to imagine such lasting power in anything to do with our notions of the fundamental nature of space, time, and matter. Similar kinds of observations have motivated so-called structural realists, who argue that mathematical formulas stay, but ontologies go. But if we take a bigger-picture view, it's not clear to me that this trend will be stable, either.

Might we not somehow settle onto a firmly entrenched and well-grounded ontology, while some fundamental change will deeply affect various observations and phenomenological laws based on the observations, or the very way we express our scientific observations and theories so that they are not mathematical in our normal sense? These are questions with no obvious answers. We have only been doing science roughly as we know it for about 400 years, or less. How can we be sure that the trends we have seen so far in science will be a good basis for predicting its future? Science is still young. It would be a dangerous fallacy to project forward the trends we have observed in the first five years of a child's life onto the rest of her life, especially if we have no idea how long that life is expected to be.

Trying to peer into the future of science, I see two drastic visions. One is that of an indefinite series of metamorphoses, each stage so fantastic that it is unimaginable to those stuck in the previous stage. We have often had this sort of changes in science, and technology. Who knows how many more such changes are left? The other vision is less exhausting but more sobering to contemplate: science may hit an unexpected dead end, with no further major discoveries forthcoming. The ground of this vision is not the self-satisfaction enjoyed by Michelson or Kelvin. No, I am imagining a future in which humanity has done all the science that we can do within our ability, and we are simply stuck in that state, knowing that there is more to learn but unable to learn it. It's an unglorious vision, comparable to imagining centuries of future Olympic Games in which no future Usain Bolt is able to run 100 metres in less than 8.5 seconds. If I were forced to make a prediction, I would have to say that I think the actual future of science would lie somewhere between my two drastic visions, or rather be a messy mixture of the two.

The Significance of the Uncertainty of the Future of Science

Having given you a sense of uncertainty about the future of science that I suffer from, I would like to conclude by considering some implications of that uncertainty.

I started out by separating out questions of foresight *in* science and foresight *about* science. Now I must note that the two questions are not entirely separable.

- Lack of foresight about science ultimately limits foresight in science.
- Limited foresight in science, in turn, significantly reduces the promise of real foresight about science.
- The future remains uncertain.
- Handling uncertainty with humility opens a new positive vision of the future, and present, of science.

The uncertainty about the staying power of scientific theories, even predictively successful ones, actually raises a serious question about the value of predictive success. There are two aspects to this. First, predictively successful theories may end up being rejected for various other reasons; contrary to Lakatos and others, predictive success is not the be-all and end-all of science and its progress. Second, predictive success itself may not last. Even for a theory that makes a string of successful novel predictions for some time, its predictive success may dry up. In fact in the Popperian, Lakatosian, and Kuhnian pictures of scientific progress this is almost taken for granted – otherwise why would scientists go over to new theories, research programmes, or paradigms? A less imaginable but possible scenario is that the predictions we used to make successfully, or the predictive methods we used to employ successfully, would somehow cease to work. This would happen if nature itself changed in relevant ways. In any case, a lack of foresight about science ultimately limits foresight in science. Limited foresight in science, in turn, significantly reduces the promise of any real foresight about the development of science. If we want to foretell the future of science, what method would we have for that, other than science itself?

All these thoughts make clear that the predictive success enjoyed by science cannot provide a sure guide to the future of science, or to the future that science seems momentarily successful in foretelling. The future remains uncertain. You knew that already before you read this far so I haven't told you anything new. The remaining question is also the obvious one: how do we prepare for an uncertain future? At one level we can only continue to live on the basis of the regularities that we do have, that have not failed us yet, while being aware that these regularities may fail at any point; this is a humble lesson that has been voiced by a range of great thinkers including Isaac Newton and David Hume. At another level, we can prepare ourselves for an uncertain future by maintaining flexibility and readiness for surprises; at the societal level, such flexibility comes from the cultivation of pluralism, about which I have spoken on other occasions.

The theme I want to emphasize here is not so much pluralism as humility, though the two go very closely together. Epistemic humility allows a view of the future of science, and its present, that avoids both of the dyspeptic visions I outlined earlier. This is not original to me, but due to Joseph Priestley, the 18th-century chemist, theologian, and political philosopher. Priestley observed: 'every discovery brings to our view many things of which we had no intimation before'. He had a wonderful image for this: 'The greater is the circle of light, the greater is the boundary of the darkness by which it is confined.' As knowledge grows, so does ignorance. But do not despair, Priestley counselled: 'But notwithstanding this, the more light we get, the more thankful we ought to be. For by this means we have the greater range for satisfactory contemplation. In time the bounds of light will be still farther extended; and from the infinity of the divine nature and the divine works, we may promise ourselves an endless progress in our investigation of them: a prospect truly sublime and glorious.'

Through his humility about human epistemic capability in comparison to the apparent inexhaustibility of the universe, Priestley was able to accept the uncertainty regarding the content of future discoveries, yet assure himself that future discoveries would continue indefinitely. Seeing the frontier of research as the boundary between a modest yet growing circle of light and an indefinitely vast darkness, Priestley was able to conceive of an unlimited expansion of knowledge in various directions that were unpredictable, yet delightful precisely for their unpredictability.

Generalizing Priestley's vision further, I would submit that in a world of insurmountable uncertainty, true foresight consists in recognizing the proper limits of our foresight.

Further Reading

Brush, Stephen G. (1989) 'Prediction and Theory Evaluation: The Case of Light Bending', *Science* 246, 1127

Descartes, René (1644) *Principles of Philosophy*

Keynes, J. M. (1921) *A Treatise on Probability*. London: MacMillan & Co.

Kuhn, Thomas (1995) *The Structure of Scientific Revolutions.* 3d ed. Chicago: University of Chicago Press

Michelson, Albert and Morley, Edward (1887) 'On the Relative Motion of the Earth and the Luminiferous Ether', *American Journal of Science* 34 (203): 333–45

Mill, John Stuart (1882) *A System of Logic.* New York: Harper & Bros.

Musgrave, Alan (1974) 'Logical versus Historical Theories of Confirmation', *British Journal for the Philosophy of Science* 25, 1–23

Popper, Karl R. Popper (1972) *Conjectures and Refutations.* 4th ed. (rev.). London: Routledge and Kegan Paul

Priestley, Joseph (1790) *Experiments and Observations on Different Kinds of Air, and Other Branches of Natural Philosophy, Connected with the Subject.* 2nd ed., 3 vols. Birmingham: Thomas Pearson

Psillos, Stathis (1999) *Scientific Realism.* London: Routledge

Scerri, Eric (2006) *The Periodic Table.* New York: Oxford University Press

Snyder, Laura J. Snyder (1996) 'Is Evidence Historical?', in Martin Curd and J. A. Cover, eds., *Philosophy of Science.* New York: W.W. Norton

Stanford, P. Kyle (2006) *Exceeding Our Grasp: Science, History, and the Problem of Unconceived Alternatives.* New York: Oxford University Press

Thomsen, William (1901), 'Clouds over the Dynamical Theory of Heat and Light', *Philosophy Magazine* (2,6) 7, 1–2

5 Music and Foresight

NICHOLAS COOK

When I was asked to give a lecture on 'music and foresight' I said yes, because one does, and then wondered what on earth I was going talk about. So I did the obvious thing and asked some of my musicologist friends what *they* would talk about in a lecture on 'music and foresight'. The second thing they all said was the same: 'Attali'. (The first thing they all said is that *they* wouldn't give a lecture on 'music and foresight'.) And when they said 'Attali', they were talking about the French economist, politician, and founding president of the European Bank for Reconstruction and Development, Jacques Attali. Among the general public, Attali is best remembered for the circumstances surrounding his resignation from the Bank, which included the £750,000 spent on replacing the red Travertine marble in the lobby of the Bank's brand-new headquarters by a paler, and in his view more aesthetically appropriate, Carrera marble.

Attali's *Noise* on Music and Social Change

But among musicologists, he is known for a book called *Noise: The Political Economy of Music*, which was originally published in 1977 and translated into English in 1985: this brought a combination of economic theory and grand cultural speculation into a discipline that was just at that time opening itself to new influences under the influence of the so-called 'New' musicologists.

So what did Jacques Attali have to say? About the marble not a lot, but he was more forthcoming on music and foresight. 'Every major social rupture has been preceded by an essential mutation in the codes of music,' he claimed, 'before the transition in political institutions

101

from divine right to political representation, the rupture had already taken place in music.' This happened, he explains, because music is 'the audible waveband of the vibrations and signs that make up society'. And more than that, 'change is inscribed in noise faster than it transforms society', and for that reason music is 'a herald of times to come'. So a principal purpose of his book is to decipher the codes and in that way reveal 'the prophecy, announced by today's music, of the potential for a new political and cultural order'.

The order to which he refers is a distinctly utopian one: it is what, in a rather idiosyncratic usage, he calls the order of 'composition', in which 'music emerges as an activity that is an end in itself'. In other words he identifies a new turn towards musical participation, and interprets it as adumbrating a broader social order that 'calls into question the distinction between worker and consumer' and hence the entire regime of monetary exchange. For Attali, 'composition' embodies the emancipation of the individual as an autonomous agent.

Noise was translated into English by no less a figure than Brian Massumi, and Frederic Jameson provided a Foreword. And in his Foreword, Jameson explained the originality of Attali's book through its inversion of the classic Marxist model according to which the economic base is reflected in the cultural superstructure. As Jameson says, Attali 'is the first to have drawn the other possible logical consequence of the "reciprocal interaction" model – namely, the possibility of a superstructure to *anticipate* historical developments, to foreshadow new social formations in a prophetic and annunciatory way'. This is a heady prospect, on the one hand empowering the artist who may contribute to social change, and on the other empowering the commentator who can read social change from musical practices. It also resonates with a small number of empirical observations that I have been able to identify: Arild Bergh writes that 'independently produced tapes of ultra-nationalistic Croatian music were on sale in stalls long before there was open war', for example, while according to Donna Buchanan, performances and recordings by Bulgarian musicians during the 1980s 'functioned not only as prominent harbingers, but as agents of political transition'. But in what way might Attali's approach provide a more general explanatory framework for such phenomena?

There is an obvious resonance between what Attali is saying and T. W. Adorno's idea that music has the capacity to 'express, in the antinomies of its own formal language, the exigency of the social condition'; it 'presents social problems through its own material and according to its own formal laws – problems which music contains within itself in the innermost cells of its technique'. It was this aspect of Adorno's approach that prompted the 'New' musicologists, and especially Susan McClary, to attempt to decipher classical scores in terms of social meaning, for example in her famous, or infamous, interpretation of the tonal drive of Beethoven's Ninth Symphony as a symbolic expression of pathological misogyny. Is this the sort of thing Attali had in mind? But then, by the time the 'New' musicologists had latched onto Adorno, sociologists had turned away from this kind of deciphering approach: as Howard Becker put it in a lecture he delivered to ethnomusicologists in 1989, 'sociologists aren't much interested in "decoding" art works, in finding the works' secret meanings as reflections of society. They prefer to see those works as the result of what a lot of people have done jointly'. In other words, you don't understand social relationships by looking in musical scores, you do it by seeing what people do in musically defined contexts – which is what Buchanan meant when she talked about Bulgarian musicians acting as agents of political transition.

So is *that* the sort of thing Attali had in mind? In truth it's hard to be sure which Attali meant, or even whether he knew which he meant. Nor is it at all clear what sort of participatory music Attali was talking about. He speaks of it as having emerged over the previous twenty years, that is up to 1977 when the book first appeared in French, and as you might expect of a French intellectual of that period, he cites composers of the European avant-garde such as Pierre Boulez, Karlheinz Stockhausen, and Luciano Berio. But they stood for an elite culture making unprecedented demands on technical skills, the opposite extreme from a participatory musical culture.

Then there are a few references to figures as disparate as John Cage, Jimi Hendrix, and The Rolling Stones. He talks at rather more length about the New York-based Jazz Composers' Orchestra Association and quotes Malcolm X on free jazz. But what he says here does not inspire confidence; Alan Stanbridge speaks of Attali's 'cursory two and a half-page analysis of a music with which he seems barely familiar, making some rather curious errors', which Stanbridge then goes on to itemise. Other than that, there is just a distinctly vague reference to 'the number of small orchestra for amateurs who play for free' and whose numbers, Attali says, have mushroomed.

In the Afterword that *she* contributed to the English version of Attali's book, however, McClary interprets Attali's 'composition' in terms of punk rock, and subsequent commentators, such as David Laderman, have followed her lead. Whether that is actually what Attali had in mind, however, is another matter. Quite apart from the fact that he doesn't mention punk, it was really only in 1977, the year the book actually came out, that punk hit the headlines: that was the year that Sid Vicious joined The Sex Pistols, and that The Clash released their debut album. Recently another, more contemporary interpretation of Attali has emerged, but I'll come to that later.

Given these fairly fundamental problems, it's probably fair to say that Attali's book got an easier ride than it might have because of the extent to which it silently builds on quite familiar thinking about music. The idea that music is a representation of society is an integral part of the discourse of free jazz, but it goes back at least to the eighteenth century, when it was in particular associated with the string quartet. In 1777, the composer and critic Johann Reichhardt referred to the quartet as 'a conversation among four persons'. But the most famous example is Goethe, who in 1829 described it as 'the most comprehensible type of instrumental music: one hears four reasonable people engaged in conversation with one another'. And it is Goethe that the contemporary composer Brian Ferneyhough is echoing when he speaks of 'the old image of four civilized people talking to each other'. If you take a page at random from the quartet literature – say the opening of Mozart's Quartet No. 14 in G major K. 387 – then you can see what is meant in the orderly succession of ideas passing from instrument to instrument. A traditional musicological description of this passage might consist of something like the following:

> A two-bar opening phrase moves from the tonic to the supertonic and is balanced by another two-bar phrase that returns through the dominant to the tonic; this pair of matched two-bar phrases leads in turn to a four-bar phrase in which a distinctive motif ascends from the viola through the second violin to the first violin.

But a description like that misses the social dimension. You get a much better sense of what is going on if you think not in terms of instruments but of the people who are playing them, or the musical roles that they are playing, in the same sense as in a role-playing game (which in a way is what chamber music is). You can't play music like this unless everyone listens to everyone else. That's because there aren't precisely standardised values of timing, intonation, or dynamics that everyone is adhering to: people aren't just playing one, two, three, four, like a metronome. Instead, all of these things are negotiated between the players in real time. In this way the music doesn't simply represent the ideal community Reichhardt, Goethe, and Ferneyhough were talking about. It enacts it, makes it present, makes it audible, and choreographs the formation of interpersonal relationships for as long as the music lasts.

Music Outside of Temporality

So the idea of music as an embodiment of society is one established trope on which Attali is drawing. A second such trope is the idea that music – though ostensibly the most evanescent of cultural practices – somehow stands outside the temporality of the everyday world. Actually this is a basic principle that has conditioned Western thinking about music for as long as you want to trace it back. In the early medieval period, the Platonic idea of earthly phenomena being copies of transcendent forms was adapted to music in a way that gave it enduring relevance to the Western 'art' tradition. Sounds were seen as copies of entities whose primary form of existence was in written form. As Sam Barrett puts it, speaking of neumes, 'notation served not simply as a pragmatic *aide-memoire*, but as a reflexive tool for disciplined knowing': more than that, it 'mirrors a higher order of being'. And this idea has survived right up to the present day in the concept of music as a form of writing, or at least an abstract entity that is embodied in or signified by writing. Seen this way, music is not sound but rather something abstract that is realised *through* sound. And since sound is inherently temporal – vibration is after all a relationship between intensity and time – this is as much as to say that music is understood as something realised *through* time but not in itself temporal. If Attali's idea that music can transcend time seems less bizarre than it really is, that is because of the engrained sense that music was never quite contained within time in the first place.

This second trope has taken many forms. Perhaps the closest to its origins in ancient Greek philosophy is its expression through shamanism, magical practices, and theories of divination; speaking of the 'natural magic' of sixteenth-century Italy, Angela Voss writes that 'The key . . . to the ordering of the cosmos, whether astronomically or musically, is . . . number . . . Number determines all things in nature and their concrete manifestation, together with all rhythms and cycles of life. Number revealed by the heavenly bodies unfolds as Time.'

This idea of music as something that is intrinsically timeless but unfolds as time also figures in the aesthetic idea of the musical work, an idea that can be traced back to the sixteenth-century music theorist Nikolaus Listenius, but is generally seen as reaching its heyday around 1800. Expressing a similar intuition in twentieth-century terms is the psychoanalytical version of the trope, the idea of a timeless, collective unconscious that music can translate into the time of experience: the general idea comes from Carl Jung, reflecting the influence of Schopenhauer, and percolating into Anton Ehrenzweig's writing on music and on art more generally. And the present-day music aesthetician Peter Kivy draws a parallel with the 'venerable theological chestnut', as he calls it, 'how does God, who is eternal and unchanging, conceive of the course of history?': the answer is that for God past and future are laid out in a kind of augmented present, and Kivy links this with the idea of genius.

Schenker and Adorno: Time, Space, and Freedom

Music theory draws on all of these ideas. The *fin-de-siècle* Viennese music theorist Heinrich Schenker, whose theories ended up through a bizarre historical tranformation as the foundation of post-war music theory in North America, gave concrete form to these tropes of number, metaphysics, and genius through an analytical method that might be characterised in terms of the notoriously inscrutable line that, in *Parsifal*, Wagner gives to Gurnemanz: 'Here time turns to space.' In a Schenkerian graph, the music is laid out as an abstract, spatial structure in two dimensions. The vertical axis represents structural depth, from foreground to background; it is a dimension of growth or differentiation. The horizontal axis represents the music's extent, which is translated into time in the act of performance. The term 'unfolding' is part of Schenkerian jargon, and you could see the music as 'unfolding' through time, rather like the unfurling of a scroll in which the future is already inscribed.

There is in short a spatial metaphor that is deeply buried at the heart of thinking about music, that bears upon our basic sense of what kind of a thing music is, and that is consequently shared by Schenker's and practically every other analytical approach. It is also, of course, the same metaphor that is built into the idea of 'foresight': the word 'forehear' exists but is little used, and some dictionaries gloss it as archaic.

This idea that music always already exists and is just unfurled or unrolled through time reached its apogee through sound recording, which could be seen as the actualisation of what aesthetics had posited only as an ideal. The 300-metre-long coiled groove of a 78-disc was originally seen as a kind of script – that was the first argument advanced for bringing recordings within copyright – in which, as with any other script, the end is present from the beginning. There even used to be an audible sign of this predetermination in the pre-echo of badly engineered LPs, literally making the future audible, in fact foreheard.

But of course the point is a much broader one. For the last fifty years there has been a widely shared view that the iterability and permanence of recordings has been reflected in a performance style that has become more and more predictable, more future-proofed, as it has converged upon a mechanical kind of perfection: 'Perfect, immaculate performance in the latest style preserves the work at the price of its definitive reification', Adorno wrote in 1938. 'It presents it as already complete from the very first note. The performance sounds like its own phonograph record.' Seen one way, the gramophone brought the concert experience into the domestic interior, creating a particularly immersive sonic analogue of the novel. Seen more pessimistically, however, this could be seen as the subordination of personal experience to a manufactured reality, in which aesthetic ideology and technology fuse into an idea of reproduction that neutralises and rationalises temporal becoming, channels it along predetermined paths even as it creates an illusory sense of freedom. In short, music turns into an instrument of the administered society.

That of course was Adorno's view, and it is here that the link between him and Attali is strongest. For Adorno, too, music might be seen as 'a rough sketch of the society under construction', and in the form of the recording it becomes not just a reflection of the commodification of experience but a mechanism for its achievement. That is Adorno's general point about the culture industry, but it is perhaps best exemplified by Muzak, a term often used generically but actually the trading name of the Muzak Corporation, or as it is now known, Muzak Holdings LLC. At one time the company focussed on the use of music for the rationalisation of production, essentially as a mechanism for mood management; nowadays, however, it focusses more on the engineering of consumption, in line with the transformation of advanced economies from a focus on manufacturing to one on service industries. Under such circumstances, borrowing Attali's broad brush, we might say that what nineteenth-century music envisaged in the aesthetic domain was realised by the end of the twentieth century in the social economy: the power of foresight once ascribed to genius has been rationalised and institutionalised in the form of consultancy services marketed as an aid to corporate planning.

Translating this back to Attali's terms, the suspicion is that we – Attali's community of self-directed individuals – are not so much the composers as the composed. There is a perhaps insoluble tension between Attali's defense of 'composition' as an exercise of individual agency and immediate pleasure, and the system of impersonal forces in which, as an economist, his thinking is grounded. Attali attempts to theorise individual freedom within a framework that militates against it. Perhaps one might say of his book what Karl Kraus said of psychoanalysis, that it is a symptom of the disease to which it presents itself as the cure. I'd also draw a parallel with the deep contradiction inherent in the old-fashioned but remarkably persistent approach to music history which on the one hand glorifies the insights of individual genius, and on the other reduces those insights to the working out – I might say unfolding – of impersonal stylistic forces.

Music, then, has been embraced within a determinedly rationalistic approach to temporal change, with time treated as a medium of presentation rather than an ontological characteristic. And the theory and aesthetics of music embody this approach in the most concentrated form. But we need to remember that music theory and aesthetics tend towards the prescription of norms rather than the description of the real experience of situated agents. Music theorists operate rather like cartographers, charting music as if seen from high above, assimilating it to deeply entrenched aesthetic and cultural values and enjoining adherence to these upon its practitioners: 'No playing the way a pedestrian might walk who gropes his way from paving stone to paving stone,' Schenker writes.

Experiencing and Creating Temporality in Music

And yet that is exactly how most of us listen to music, hearing one thing after another, following its phenomenology from one moment to the next, taking pleasure in precisely those irreducibly temporal transitions that disappear from traditional analytical graphs. When we listen to music, time is of the essence. Whereas analysis sucks time out of music, reduces it to the known, the foreseen, it is intrinsic to the enjoyment of music that the pleasure of expectation is balanced by the pleasure of expectations being controverted, or even the pleasure of having no expectations at all.

Even when we know the music backwards, as the revealing phrase has it, what we experience in the real time of listening exceeds what we remember, overwrites and in that sense erases it. And the same applies to what we foresee, in those musical genres that are designed to give you the pleasure of knowing what's coming next. Repetition is an almost universal musical phenomenon that theorists have always found baffling because they have adopted a conceptual framework that makes it baffling. The point is that what you hear is never quite what you foresaw; there is always a residue of astonishment (or at least if there isn't, it's time to stop listening). And the result is you always seem to be listening for the first time, even when you know the music backwards. This is one of the ways in which music helps you to be sure you are alive.

But it's not just the fact of music's temporality, it's its nature. Conceptualisations of music dominated by space and structure have given rise to the idea that music exists in time in the same sense that humans exist in air or fish in water. It's a neutral kind of time, a transparent medium, a time that ticks away in evenly spaced increments. That's the conception of time that has dominated performance, both classical and popular, since the Second World War, the time within which you count 'one, two, three' and fit rhythms within the beat.

But time didn't always pass that way. We know that from the performances you hear in the earliest sound recordings or piano rolls. In the playing of the once famous pianist and composer Carl Reinecke, for example, who was born in 1824 and is possibly the oldest musician whose playing we can still hear, time isn't a transparent medium. Rather it is a dimension of the music. It's partly to do with distinguishing different characters in the music, and partly to do with bringing out the music's gestural quality, but it's also that Reinecke slows down when the music becomes denser – whether in terms of melody, harmony, texture, or emotional intensity – and speeds up whenever it becomes less so. The result of all this is that he is constantly changing speed, now lingering and now rushing ahead, and the same applies to other pianists of the early twentieth century, for instance the once famous Beethoven interpreter and composer Eugen d'Albert, who died in 1932.

But as the new conception of time developed during the interwar period and then took over after 1945, the way of playing captured on early recordings became incomprehensible. Writing in 1965, Harold Schonberg wrote that d'Albert's recordings 'cause nothing but embarrassment ... the playing is inexplicable, full of ... distorted rhythms', while the author of a liner note for the modern reissue of one of those recordings, whom you might expect to be on the pianist's side, refers to his 'strange rhythmic lapses'.

Contrary to Schonberg, d'Albert's playing isn't inexplicable at all: it's simply that time went differently a century ago. And this new concept of time coincided with a change in how musicians thought about musical texts, or at any rate, how they treated them in practice. Pianists like Reinecke and d'Albert respected the classics, but they expressed this respect by treating classical works not so much as texts to be reproduced but rather as scripts to be produced in a theatrical sense. They had a healthy scepticism about what could be grasped in notation or in theoretical language, and consequently a healthy belief in their own importance as interpreters.

The difference between them and modern performers lies in the new and more literal concept of fidelity to the score that became dominant after the War: it is this, and not just the influence of the gramophone, that Adorno heard and complained about in 1938. There is a comparison with what the famous post-war musicologist Carl Dahlhaus referred to 'a "popular Platonic" aesthetics ... which tends to fasten on the paraphrases that are set up "around" a thing, and treats them as the ideas that lie "behind" the concrete phenomena and manifest themselves through them'. You might say that musical scores were hypostasised in just the same way, and not only in terms of musicology but also of performance. Perhaps the effect has been to diminish the sense of astonishment to which I referred.

The Foreseen and the Unforeseen

Composition is a particularly revealing example of this tension between literalism and interpretation, the foreseen and the unforeseen. The kind of music-historical writing I spoke of earlier, in which style changes are understood as the unfolding of impersonal forces, brings the same kind of deterministic approach to bear upon the compositional process. Twentieth-century composition has been characterised by an unprecedented reliance on explicit, systematic methods.

Serialism, which prescribes note-to-note successions and in its later forms also prescribed the organisation of rhythms, dynamics, registers, or timbres, was frequently justified by its advocates as a form of rationalisation, ensuring coherence and the consistent and predictable handling of the various musical parameters. But here appearances are deceptive. In reality serialism worked more like surrealist automatic writing, fixing certain elements while leaving others completely undetermined and so promoting a kind of radically improvisatory approach. Or by proposing something, the system elicited a response from the composer, quite possibly one of rejection, resulting in a kind of open-ended dialogue between composer and system.

I can illustrate this by reference to a contemporary American composer, Roger Reynolds, who makes extensive use of mathematical series in his compositions. But his employment of this apparently objective technique is in fact highly subjective: 'although a very considerable amount of time goes into exploring alternative rows,' he writes, 'the process by which one possibility is discarded and another is more extensively searched remains for me a very personal one, resistant to objectification'.

Systems like this do two things. First, they prevent you from following the line of least resistance, doing what you usually do, and so imitating yourself, falling into your own clichés. Second, they elicit a dialogue through which, in reacting to what the system proposes, you discover what it is that you really want. In short, the point of the system is not to bring about the foreseen. It is to guarantee a compositional outcome that was *not* foreseen. It is almost as if the more deterministic the compositional process, the less predictable its consequences.

I nearly said that by rejecting what the system proposes, you discover what it is that you really wanted all along. That is how it seems. Music is not an art of foresight, but one that creates illusory impressions of foresight, holding out the promise of a rationalised utopia where everything is intended. But look further, as I have suggested, and music is an art concerned at most with the mitigation, and sometimes with the celebration, of the unpredictable. And actually an argument might be made that, despite the best efforts of music theory and aesthetics to rationalise music, to contain or conceal its radical temporality, music is in reality a field *especially* resistant to foresight.

Despite hoary claims that music is a universal language, it is as closely tied to time as it is to place: that is the point of the Armstrong and Miller sketch in which an early nineteenth-century social gathering is disrupted as the pianist repeatedly spices up his decorously period playing with stylistic elements from twentieth-century American popular music. Presumably it is just because of the closeness of this tie – because music is in so high a degree a fashionable art – that the best efforts to predict the music of the future fail so spectacularly. *Space Mutiny*, a South African film from 1988, is set at some unspecified time in the future, aboard a space ship on a multi-generational journey towards new worlds to colonise. At one point there is a rave scene: despite the film-makers' sincere attempts to envision future female fashions through liberal use of Spandex, not to mention their innovative approaches to hula-hoop dancing, the music could hardly betray its 1980s origins more transparently. And Klingon opera, as depicted in a 1991 episode from *Star Trek* set in the year 2368, turns out to be made up from elements drawn from a combination of nineteenth-century Western opera and all-purpose barbarism in almost exactly the same way that the orientalising composers of that century depicted the exotic cultures of their day. It seems that, despite our claims to grasp musical developments within overarching style histories – or the bathos of Franz Liszt's 'music of the future' and Ornette Coleman's 'The shape of jazz to come' – we can't see what is coming even a few years down the line.

Actually it's pretty impossible to foresee what is going to happen even a few weeks down the line. That's why the pop record industry's practice was always to produce huge numbers of discs, the vast majority of which disappear without trace – but with the few that succeeded generating the high level of profits that the industry enjoyed during its long heyday. Clearly record companies would not have worked this way if they had been capable of predicting hits with any degree of reliability.

Over the last decade, developments in Music Information Retrieval (MIR)—a combination of digital signal processing, cognitive psychology, and musicology oriented towards commercial applications – have brought this prospect closer. Huge investment has been put into music-preference systems, which are designed to predict whether, if you like songs A and B, you will like song C, generally on the basis of some kind of multi-dimensional analysis of song characteristics. If you can do that, shouldn't you equally well be able to predict hits? A recent project called ScoreAHit, based at the University of Bristol, uses a mathematical equation based on the weighting of a number of salient characteristics, and it achieves a 60 per cent success rate in identifying the songs from the last fifty years that became hits.

This kind of retrospective prophecy sounds impressive, and indeed it *is* impressive, but not quite as impressive as it sounds, because the dataset was designed to contain equal numbers of hits and flops. In other words, sticking in a pin at random should achieve a 50 per cent success rate, and ScoreAHit adds a further 10 per cent. One reason it's not better is the same one that makes predicting the weather so hard, even with the fastest computers: too many interacting elements. But in the case of pop music there is a further basic problem: ScoreAHit are simply working from the audio, whereas there are any number of other factors – from record industry politics to the strength of an established artist's fan base to sex appeal – that feed into what turns a song into a hit, not to mention that underestimated determinant of world history, luck.

Perhaps the most spectacular example of lack of foresight that music can provide, however, is furnished by the record business. In the course of a century it went from being a focus of cutting-edge, even visionary business thinking to one that almost destroyed itself through a doomed attempt to resist technological and behavioural change. From the earliest years of the twentieth century, the Gramophone Company had a global strategy that involved developing branches across the world whose role wasn't simply to source products for import to the West, in the manner of raw materials or luxury goods, but rather to create and service local demand. The record business began to assume its modern form in the 1930s, with a series of mergers that eventually left most of the industry in the hands of a small number of global majors – it's now down to four – and with decades of profitability resulting from the successive technologies that led consumers to repurchase their collections over and over again (acoustic 78s, mechanical 78s, LPs, stereo LPs, CDs). This resulted in a business model that was firmly grounded on the selling of a musical product, preferably many times over.

That's why the digital technologies that enabled unlimited, cost-free copying and format shifting without loss of quality struck at the heart of the record industry's business model. As everyone knows, the majors responded through successive attempts to prevent technological innovation and internet-based dissemination. Nemesis followed in a manner worthy of Greek tragedy: while the record companies put their efforts into suing astonishing numbers of their own customers for copyright infringement, Apple and Amazon stepped in and established themselves as the leading players in music dissemination via the internet. At the same time, other players developed a host of related service industries, as exemplified by Spotify and last.fm. But the largest of these service industries is actually the live music business, that's to say the promotion of concerts, long seen by the record companies as ancillary to record sales, but since the mid-1990s the most profitable sector of the music industry.

The timing is revealing: sales figures show that records were already losing out to live music in 1996, which is to say five years *before* peer-to-peer file sharing became a major issue – which completely undermines the case on which the majors based their demands for the tightening of legal restrictions on the internet. Despite this, the record industry succeeded in convincing politicians that it *is* the music business: like the failing Detroit motor industry, its greatest success was in political lobbying.

What is telling about all this is the extent to which record business thinking was moulded – and boxed in – by the same product-based thinking that I was talking about earlier on: the whole trend of the conceptualisation of music that I traced back to Plato is to emphasise product at the expense of process, or as musicologists put it, music at the expense of musicking. That is perhaps how it came about that the new players who grabbed the music business while the majors were fighting yesterday's battles came from backgrounds other than music: the founders of Spotify and last.fm came from deejaying, videogames, computer engineering, and internet marketing.

From Products to Prosumption

Perhaps the principal task of my generation of musicologists has been to broaden the approach of a discipline that historically treated music in the way the music industry did, as a product, a commodity, instead of seeing it as a social practice, in Becker's words 'what a lot of people have done jointly'. What brought the old order to an end, certainly in commercial terms and perhaps in aesthetic and theoretical terms too, was not so much new technologies as such, but rather the social practices of production and consumption – or prosumption – that new technologies made possible. I'm talking about Henry Jenkins's 'convergence culture', which he styles as a revival of traditional American folk values enabled by web 2.0: as he says, consumers have become participants, as illustrated through such burgeoning areas as internet-based fanfiction and the participatory video culture based around YouTube.

Again, you find the same strain of digital nostalgia in Lawrence Lessig's 'remix culture', which he also refers to as 'Read/Write' or RW culture: in essence this is the same idea as Jenkins's – even down to the American folk revival aspect – but now with an emphasis on the 'hybrid economy' that could result from a more permissive copyright regime.

(Lessig is a co-founder of Creative Commons.) That explains the subtitle of his book, but he takes his main title from the practices of hip-hop and mash-up that have been the principal battleground between a music industry committed to perpetuating old business models and the new breed of prosumer.

Similar to Jenkins's convergence culture and Lessig's remix culture is what Aram Sinnreich calls 'configurable culture'. Sinnreich uses this term to refer to genres like hip hop and mashup, along with the practices of prosumption, dissemination, and exchange that have built up around them. But it's easiest to define what he means in terms of what he sees it as opposed to: what he calls the 'modern discursive framework'. This includes the traditional dimensions of musical thinking and practice that I have been talking about, among them the ideas of music as a form of writing, as a spatially extended object, and as a commodity; as Sinnreich puts it, the basic premise of this framework is 'that music exists both prior to and independently of its expression'. It also includes the established music business. But Sinnreich's modern discursive framework extends wider than that, taking in traditional conceptions of genius, originality, and creativity; the regime of copyright that is built on that concept of creativity; and the entire industrial and economic system that is in turn built on copyright. (Here his thinking is similar to Lessig's, and indeed Lessig's 'Read/Only' or RO culture effectively maps onto Sinnreich's modern discursive framework.) Sinnreich's claim, then, is that the new opportunities the culture of digital music creates for the generation of wealth, and the threats it presents to traditional business models, offer within a realm of pure signs the same opportunities and threats that will soon present themselves across the commodity economy and society in general. 'The resolution of this crisis', he says, 'will help to determine the organizing principles of postindustrial society for years or perhaps for centuries to come.'

Similar claims have been made by other analysts of digital culture. In his book *Watching YouTube*, Michael Strangelove quotes Nancy Byam's claim that 'fandom is a harbinger of cultural phenomena to come', which was based on a study of Swedish online communities, and adds that this 'applies to the entire YouTube community of amateur videographers'. Again, Tom Boellstorff writes in his book on Second

Life that virtual worlds have revealed new possibilities of social control, but at the same time 'they also indicate how forms of resistance could challenge, reshape, or work around these encodings of sociality'. Music shares many of the characteristics of virtual worlds, including the fact that – as Boelstorff also says of Second Life – both 'draw upon many elements of actual-world sociality, but ... reconfigure these elements in unforeseen ways'.

Seen this way, Sinnreich's configurable culture is 'a staging ground for new social ideas'; it allows for the emergence of new social forms or business practices – and of modes of resistance to them – with far less investment and drag, and hence far more quickly, than in the world of physical commodities. No wonder then that Sinnreich's book is littered with such terms as 'harbinger', 'foretaste', 'presage', 'prefigure', and 'portend'. In this way he brings a degree of specificity and plausibility to Attali's suggestive but ineluctably vague idea of music as foresight, as well as to Attali's utopian approach to musical participation.

Of course, unless he was gifted with foresight of a different order, Attali was not thinking about Sinnreich's configurable culture, which is firmly tied to digital technology. While there are examples of some of the individual elements of configurable culture in the past, Sinnreich says, 'the configurable media experiences of the present day clearly outnumber, overpower, and outpace any of these examples by orders of magnitude': in this way they amount to a fundamentally new paradigm. Here Sinnreich differs from Lessig, who goes out of his way to emphasise that what he calls remix or RW culture has always existed: a hundred years ago, he says, 'the creativity was performance. The selection and arrangement expressed the creative ability of the singers'. Sinnreich specifically rejects this claim: 'I must disagree with any claims of continuity between past and present cultural practices.' But if you look at sources like the *Hofmeister Monatsberichte*, monthly listings of new sheet music that were issued by the Leipzig-based publisher of that name from 1829 until well into the twentieth century, the classics as we know them almost disappear under a flood of arrangements for every conceivable combination of instrumental forces (and some inconceivable ones, such as an ensemble consisting of physharmonica, harmonic flute, and accordion). What are these if not period remixes, the nineteenth-century version of configurable culture?

It's at moments like this that you realise how little connection there is between classical musical culture as we know it, and classical musical culture as it actually was. As Lessig says, RO culture – the culture epitomised by the gramophone, with its ability to play back but not to record – has been just one element within traditional cultural systems. And the assumption that RO culture is the only culture to be taken seriously is a phenomenon of the second half of the twentieth century, brought about through the combination of technology, aesthetic ideology, education, and post-war arts policies.

It's ironic: it took digital technology and web 2.0 to make me really think about the extent to which the practice of music is and always has been a predominantly participatory culture, with RO culture representing the icing on a much larger cake. I arrived at this obvious truth via a project on the different versions and remakes of the 'Bohemian Rhapsody' video. Planning to talk about perhaps a dozen professionally made versions ranging from commercials for global corporations to the Muppets, I discovered the existence of literally thousands of home-made versions exploiting every imaginable iconography: there is a whole repertory of Star Trek versions, for example, including one that draws on the performance practices of Klingon opera. And all this is surrounded by the lively, if not invariably edifying, discourse of YouTube commentary.

Now I spoke of musicologists of my generation rethinking music in terms of not product but process, not music but musicking. But the RO paradigm is too deeply entrenched in musicology for participatory culture to have percolated to the core of the discipline. I found myself faced with an overwhelming plurality of material for which my musicological training simply hadn't prepared me. Yet work on nineteenth-century musical culture presents you with precisely the same problem as YouTube culture, at least if you are to come to terms with the hardly less overwhelming plurality of performance practices whose traces are gathered in the *Hofmeister Monatsberichte* – as opposed to what musicologists normally do, which is to cherry pick the canonic repertory and so end up with the highly skewed interpretation of that culture that still passes for music history.

If then contemporary digital cultures can act in one respect as a harbinger of future trends, as Sinnreich says, then equally they can act in another respect as a means of gaining greater insight into the past: the participatory cultures that Jenkins, Lessig, and Sinnreich describe are in this way a continuation by other means of traditional participatory practices, which is the point of Jenkins's and Lessig's nostalgic invocation of American folk culture. More generally, music may give us insight into certain aspects of life experience in the past, for example in the ways scores choreograph corporeal postures, gaits, and gestures (think of Domenico Scarlatti's sonatas), or in the way they equally choreograph interpersonal or social relationships among performers (as we saw in the Mozart quartet) – not to mention the different kind of time you hear on those old recordings. Equally, it could be that marginal or apparently erratic aspects of present-day digital music adumbrate what in the future will become normal experience. Yet I wonder whether this actually amounts to more than a roundabout and rather misleading way of saying that music gives us insight into the present – and therefore, indirectly and to a strictly limited degree, into what led to and will emerge from that present.

Sinnreich writes that 'music possesses a unique power to reflect, transmit, and amplify ... the social imaginary', an observation that resonates with Attali's claim that music 'is a tremendously privileged site for the analysis and revelation of new forms in our society'. But new forms are still present forms. And what is necessary in order to capitalise on music's presentation of them is an approach that is oriented towards present experience. The established, spatialised aesthetics that sees music as a commodity separates it from present-tense experience, and constitutes listening as an engagement with works which by definition already exist, whether by virtue of being old or being eternal. At the same time it encloses music within the concert hall in the same way that ancient artefacts are enclosed within museum cases, or ritualises its consumption within the home through the use of headphones in a darkened room. And critically, it represents this as the only way of listening to music, or at least the only creditable way, whereas in reality music has always been listened to in other ways.

Consider music heard through earbuds on the underground and in this way woven into the fabric of everyday life. In this context music is used not

119

just as a means of mitigating uncertainty, as I said earlier, but of managing it, of keeping a grip on the conflicting demands and incipient chaos of everyday life, and in so doing maintaining, negotiating, and presenting the self. (Desert Isand Discs tells us that people have been doing that with music for at least the past seventy years.) Music becomes a means of self-optimisation, of taking control of one's experience, of personalising the passage of time: it can even serve as a mode of resistance to the administered mood management purveyed by the likes of Muzak Holdings LLC. It shapes the passage from one moment to the next, regulates the orderly transformation of the future into the present, in that sense domesticates the future. In short, music serves as a medium through which we both heighten our experience and reflect on the dynamics of the present moment.

This idea of music as a means of optimisation extends into the social domain. Research into music's adaptive potential (in which the Cambridge Faculty of Music is playing a leading role) shows the extent to which entrainment – our ability to identify both mentally and physically with a predictable beat, and in collectively identifying with it to identify with one another – serves as a means of social coordination, including the management of affective experience and the development of empathy. Here shared experience becomes a vehicle for social relationship.

For musicologists as for the rest of us, music discloses and nuances both the passage of time and our ability to enter into social relationships with unrivalled finesse, transforming them into sound and so capitalising on the extraordinary sensitivity to nuance that characterises human hearing. That's where Attali and Sinnreich are right. But it follows from this that music's core power lies in the enhancement and understanding of present experience. It may hint at who we were in the past and who we might be in the future, but more importantly it is one of the ways we are who we are, right now.

Further Reading

Adorno, Theodor (2002) *Essays on Music*, ed. Richard Leppert. Berkeley: University of California Press.

Archbold, Paul (2011) 'Performing complexity: a pedagogical resource tracing the Arditti Quartet's preparations for the Première of Brian

Ferneyhough Sixth String Quartet', http://itunes.apple.com/gb/itunes-u/arditti-quartet/id441504831

Attali, Jacques (1985) *Noise: The Political Economy of Music*, trans. Brian Massumi. Manchester: Manchester University Press

Barrett, Sam (2008) 'Reflections on music writing: coming to terms with gain and loss in early medieval Latin song', in Andreas Haug and Andreas Dorschel, eds., *Vom Preis des Fortschritts: Gewinn und Verlust in der Musikgeschichte*. Vienna: Universal Edition, 89–109

Becker, Howard (1989) 'Ethnomusicology and sociology: a letter to Charles Seeger', *Ethnomusicology* 33, 275–85

Bergh, Arild (2011) 'Emotions in motion: transforming conflict and music', in Irène Deliège and Jane W. Davidson, eds., *Music and the Mind: Essays in Honour of John Sloboda*. Oxford: Oxford University Press, 363–78

Boellstorff, Tom (2008) *Coming of Age in Second Life: An Anthropologist Explores the Virtually Human*. Princeton: Princeton University Press

Brennan, Matt (2010) 'Constructing a rough account of British concert promotion history', *Journal of the International Association for the Study of Popular Music* 1/1, 4–13

Buchanan, Donna (2005) *Performing Democracy: Bulgarian Music and Musicians in Transition*. Chicago: University of Chicago Press

Cook, Nicholas (forthcoming) 'Video cultures: "Bohemian Rhapsody", "Wayne's World" and beyond', in Joshua Walden, ed., *Representation in Western Music*. Cambridge: Cambridge University Press

Dahlhaus, Carl (1991) *Ludwig van Beethoven: Approaches to his Music*, trans. Mary Whittall. Oxford: Oxford University Press

Jenkins, Henry (2006) *Convergence Culture: Where Old and New Media Collide*. New York: New York University Press

Kivy, Peter (1993) *The Fine Art of Repetition: Essays in the Philosophy of Music*. Cambridge: Cambridge University Press

Laderman, David (2010) *Punk Slash! Musicals: Tracking Slip-Sync on Film*. Austin: University of Texas Press

Lessig, Lawrence (2008) *Remix: Making Art and Commerce Thrive in the Hybrd Economy*. London: Bloomsbury

McClary, Susan (1991) *Feminine Endings: Music Gender and Sexuality*. Minneapolis: University of Minnesota Press

Mazzola, Guerino, in collaboration with Sara Cowan, I-Yi Pan, James Holdman, Cory Renbarger, Lisa Rhoades, Florian Thalmann, and Nikolai Zielinski (2011) *Musical Performance. A Comprehensive Approach: Theory, Analytical Tools, and Case Studies*. Heidelberg: Springer

Reynolds, Roger (2002) *Form and Method: Composing Music*, ed. Stephen McAdams. New York: Routledge

Schenker, Heinrich (2000). *The Art of Performance*, ed. Heribert Esser, trans. Irene Schreier Scott. New York: Oxford University Press

Sinnreich, Aram (2010) *Mashed Up: Music, Technology, and the Rise of Configurable Culture*. Amherst: University of Massachusetts Press

Small, Christopher (1998) *Musicking: The Meanings of Performing and Listening*. Middletown: Wesleyan University Press

Stanbridge, Alan (2008) 'From the Margins to the Mainstream: Jazz, Social Relations, and Discourses of Value', *Critical Studies in Improvisation* 4, https://journal.lib.uoguelph.ca/index.php/csieci/article/view/361/960

Strangelove, Michael (2010) *Watching YouTube: Extraordinary Videos by Ordinary People*. Toronto: University of Toronto Press.

Vial, Stephanie (2008) *The Art of Musical Phrasing in the Eighteenth Century: Punctuating the Classical 'Period'*. Rochester: University of Rochester Press.

Voss, Angela (1998) 'The Music of the Spheres; Marsilio Ficino and Renaissance harmonia', *Culture and Cosmos* 2/2, http://www.rvrcd.com/essays/The%20Music%20of%20the%20Spheres.pdf

6 Foreseeing Space Weather

JIM WILD

For as long as humans have walked the surface of the Earth, a curious astronomical coincidence has periodically offered observers tantalising glimpses into the true nature of our nearest star. Despite being nearly four hundred times closer to the Earth than the Sun, the Moon is roughly four hundred times smaller in diameter, meaning that their angular sizes are almost identical when viewed from the Earth. The total solar eclipses that arise when the two bodies align in the sky reveal the solar corona – the solar atmosphere that is normally hidden from view by the dazzling glare of the solar disc (Figure 6.1).

Counter-intuitively, modern spectroscopic measurements have revealed that the corona is much hotter than the surface of the Sun, with coronal temperatures peaking in the region of two million kelvin, compared to around six thousand kelvin at the visible surface. Although the exact coronal heating mechanism remains the focus of current research, the transfer of energy from Sun's massive magnetic field is now understood to play a crucial role. The striations and loops in the corona visible during a total solar eclipse reveal the structure of solar magnetic field embedded with the ionised gas, or plasma, that makes up the corona.

Prior to the advent of modern astronomical techniques, the brief and occasional glimpses of the corona offered by total solar eclipses were not the only evidence that he Sun was not a uniform, unchanging object. In the seventeenth century, Galileo turned his groundbreaking tele-scopes towards the Sun and (after taking suitable precautions) observed dark features that moved across the solar disc, some waxing and waning over a period of days. Regular observations made over a number of weeks revealed that the motion of these sunspots was consistent with

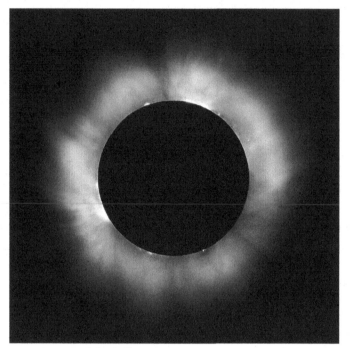

FIGURE 6.1 Solar Eclipse, 1999

surface features on a spherical body rotating roughly once every twenty-seven days. Crucially, these observations revealed that the Sun was an active and dynamic body.

By the mid-nineteenth century, a link between solar activity and geomagnetic disturbances here on Earth was beginning to be discussed. A solar flare in the vicinity of a large sunspot group reported by English astronomer Richard Carrington in September 1859 was shortly followed by a massive geomagnetic storm, the consequences of which were felt all over the globe. Unprecedented displays of the auroral borealis were seen as far south as the Caribbean and intense magnetic disturbances interfered with the operation of the emerging communication technology of the day, the electric telegraph. Although this disturbance, and many more numerous but less intense geomagnetic storms, appeared to follow some indications of solar activity, the mechanism by which energy propagated from the Sun to the Earth was not clear. The Sun's overall thermal and optimal emissions barely changed during a solar flare and the vast distance between the Sun and Earth would mean that the increases in radiative output required to provide the energy to drive geomagnetic storms should have been easily detectable, but were not observed.

A Nineteenth-century Space Weather Forecast

However, as the nineteenth century drew to a close, a Norwegian polar expedition undertook measurements of the magnetic field at the group to determine the global pattern of electric currents in the polar region. In the following years, the Norwegian scientist Kristian Birkeland developed laboratory experiments that used magnetic fields to control electron beams fired towards a magnetised model of the Earth. Using his *Terrella* devices, he showed that the resulting electric currents were guided by the magnetic fields towards the polar regions and produced light. These experiments led to the suggestion that energetic electrons, emitted by sunspots in the direction of the Earth, are guided towards the magnetic poles by the geomagnetic field where they produce visible aurora. However, the theory that energy was transferred from the Sun to the Earth not just by electromagnetic radiation, but also via particles with mass travelling rapidly through interplanetary space, was to remain untested until the dawn of the space age.

Space-age Foresight

Shortly after its launch in early 1958, the USA's first successful satellite, Explorer 1, started returning data from a cosmic ray detector designed by physicist Professor James Van Allen from the State University of Iowa. The measurements suggested that the Earth was surround by regions filled with high-energy, electrically charged particles, prompting one of van Allen's colleagues to exclaim 'My God! Space is radioactive!' The discovery of the Earth's radiation belts, still commonly referred to as the 'Van Allen belts', was the first of many landmark discoveries that followed the beginning of the space age. In the half-century that has followed, Earth-orbiting satellites have mapped out the space environment surrounding our planet, robotic probes have sampled the gulf interplanetary space and explored the planets of our solar system, and space-based solar telescopes, free of the Earth's atmosphere, have provided unprecedented views of the Sun.

We now understand that the solar corona is expanding continuously, forming the solar wind. This magnetised plasma, predominantly consists of ionised hydrogen originating from the solar atmosphere. It flows constantly outwards through the solar system until it is confined somewhere beyond the orbit of Pluto by the pressure of the interstellar medium. The solar wind flows supersonically past the Earth (typically at 300–500 kilometres per second) and due to its electrical conductivity, it carries with it the remnants of the Sun's magnetic field known as the interplanetary magnetic field (IMF). The Earth's relatively strong magnetic field acts as an obstacle to the solar wind, slowing and diverting the flow around the planet, forming a cavity in the solar wind known as the magnetosphere (Figure 6.2).

Interactions between the solar wind and the magnetosphere are primarily controlled by the relative orientation of the Earth's magnetic field (which is essentially fixed on timescales less than years) and

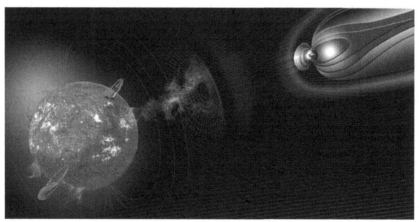

FIGURE 6.2 Magnetosphere

the IMF embedded in the solar wind (which is highly variable on timescales of minutes). The resulting solar wind–magnetosphere coupling stirs convective plasma flow in the magnetosphere and drives an extensive and complex set of electrical currents, some of which close through the ionosphere (the electrically conductive upper portion of the Earth's atmosphere). Changes in the effectiveness of this coupling are responsible for the time-varying nature of many common phenomena in the near-Earth space environment, including dynamic auroral displays and the filling/draining of the energetic particles populations within the radiation belts.

In addition to continuously emitting the magnetised solar wind, the Sun can exhibit sudden, violent behaviour. Explosions at the solar surface emit massive quantities of energy ($\sim 10^{25}$ joule) in the form of radiation spanning the electromagnetic spectrum from radio waves to X-rays. Travelling at the speed of light, it takes $\sim$8 minutes for a solar flare's radiation to propagate to Earth. Billion-tonne clouds of magnetised plasma known as coronal mass ejections (CMEs) often accompany solar flares. They can be flung out from the corona in almost any direction at a million miles per hour and if, by chance, a CME is launched Earthward, it will take typically between 1 and 3 days to traverse the space between the Sun and our planet, constantly evolving as it interacts with the ambient solar wind flow.

Hazards of Space Weather

The biological hazards at the Earth's surface due to this solar activity are minimal. The enhanced flux of X-ray and extreme ultra-violet radiation during a solar flare increases the ionisation rate of gases in the upper layer of the dayside ionosphere, enhancing the electrical conductivity and strengthening pre-existing ionospheric electric currents. These enhanced currents exert a magnetic influence at the ground, but the strength of such disturbances is typically a few tens of nanotesla (nT), compared to a surface field strength $\sim$30,000 nT at mid-latitudes. When CMEs impact the Earth's magnetosphere, their energy and electrical plasma enhance existing magnetospheric currents resulting in disturbances that can last for several days, a phenomenon known as a geomagnetic storm. The enhanced magnetospheric currents can cause large magnetic variations at the ground (up to ~10% of the Earth's surface field strength) but without the aid of magnetic field-sensing equipment an observer on the ground might be oblivious to the solar assault on the Earth's magnetic field.

The most obvious symptoms of such geomagnetic disturbances are increased auroral activity. The aurora borealis (and their southern counterpart, the aurora australis) are a consequence of energetic electrically charged particles from the space environment precipitating into the upper atmosphere. In essence, the magnetosphere behaves like a natural particle accelerator, drawing energy from the solar wind energy and using it to accelerate magnetospheric plasma, some of which leaks into our atmosphere. The most common auroral emissions, the characteristic green and red colours, are produced by the bombardment of atomic oxygen at altitudes of around 100–250 kilometres by precipitating electrons. The configuration of the inner magnetosphere is such that aurorae are typically observed in a pair of crown-like ovals surrounding the Earth's magnetic poles (Figure 6.3), but during geomagnetically disturbed intervals, such as those sparked by a CME impacting the Earth's magnetosphere, the auroral ovals can expand equatorward to much lower latitudes.

The complex pathway that electromagnetically links the Sun to the Earth has been the focus of an active international community of solar-terrestrial physicists since space-age technology allowed direct access to the space environment. Auroral activity serves as a useful barometer of geomagnetic activity and is extensively studied using optical, magnetic and radar experiments from the ground, sounding rockets and orbiting satellites. The science is now reaching maturity, with much of the solar-terrestrial environment mapped and characterised, but with several significant questions remaining in our understanding of the relevant physical mechanisms.

Now, at the beginning of the twenty-first century, the state-of-the-art is such that comparison can be made between the study of the space environment and the objectives and methodologies of terrestrial meteorology. Indeed, the World Meteorological Organisations takes an interest in the subject, defining 'space weather' as encompassing 'conditions and processes occurring in space, including on the sun, in the magnetosphere,

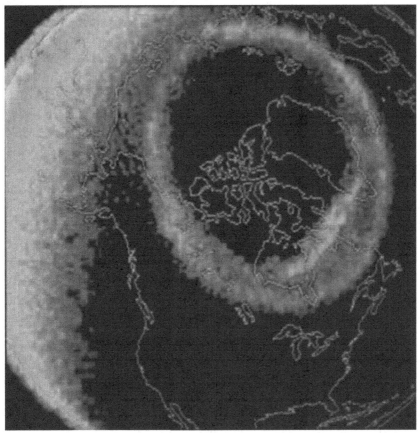

FIGURE 6.3 Auroral Oval

ionosphere and thermosphere, which have the potential to affect the near-Earth environment'. So what of foresight? The Sun's electromagnetic connection to our planet has remained broadly unchanged for billions of years, so why should we be looking towards the future? What benefit is there to understanding the impact of space weather upon the Earth?

Why Forecast Space Weather?

Our nearest planetary neighbour provides one answer. In many respects, Mars is rather like the Earth: the radius of its orbit around the Sun is only slightly (approximately 50%) larger than the Earth's and its diameter is only slightly (approximately 50%) smaller. Like the Earth it is a rocky body, and the like the Earth it probably has a metallic core. Compared to the truly alien gas giants in the outer solar system, Mars looks very similar to our own planet. But in other important respects, there are stark differences: Mars is a cold, dry world shrouded in a thin atmosphere (the surface atmospheric pressure at Mars is almost two hundred times lower than at Earth). So why are these neighbouring worlds so different? The reason is partly because of space weather.

When the Earth and Mars were formed in the early solar system, both planets comprised a mixture of rocky and metallic material, with the denser metals ultimately settling in the centres of the planets, forming cores of nickel and iron. In the Earth's case, enough of the heat of formation (and the heat produced by the decay of radioactive elements) remains to this day for the metallic core to remain molten. Currents flowing in the liquid core give rise to a dynamo action, generating the strong magnetic field that provides a force field to hold the solar wind at bay. But at Mars, the planet's smaller size meant that it contained less heat after formation and that the planet cooled at a much faster rate. The molten core probably froze and solidified about four billion years ago. Without a dynamo at its core, Mars has not had a strong magnetic field to divert the solar wind flow around the planet and the solar wind has had direct access to the Martian atmosphere. Over billions of years, this has contributed to the loss of the red planet's atmosphere to interplanetary space.

Could this happen on Earth? It's unlikely that the Earth's dynamo is going to shut down permanently any time soon, but information about the Earth's magnetic field pattern found in volcanic rocks indicates that the terrestrial magnetic field reverses four or five times every one million years. The mechanism and timescale of these reversals is not clear, but it is possible that they involve a period when the Earth's magnetic field is significantly weakened lasting several thousands of year. But over this short (geologically speaking) timescale, the impact on the Earth's atmosphere should not be significant. Our atmosphere will continue to shield us from harmful space radiation, even if our atmosphere's magnetic shield is temporarily weakened, causing the rate of atmospheric loss to the space environment to increase temporarily. The geological record reveals many such reversals in the past and life on Earth has continued to proliferate.

However, although the Earth's atmosphere and magnetosphere shield the surface of the Earth from the biological hazards of space weather, the same is not true for some of the technologies developed in the last fifty years. The natural processes in the space environment remain the same as ever, but our adoption and reliance on vulnerable technology has increased the space weather hazard. As a consequence, foresight is essential. There is a need to understand how our high-tech infrastructure will respond to space weather and to assess how bad space weather can get. In this respect, to understand the risks ahead, we need to look back.

By examining historical magnetic observatory records, the geomagnetic disturbance that followed the solar flare observed by Carrington in September 1859 has been identified as the largest geomagnetic storm on record. The magnetic disturbances during this episode, now known as the 'Carrington event', were so intense that electric currents were induced in the copper lines of the Boston to Portland telegraph strong enough that the operators were able to continue to use their sets to transmit messages even when the power supplies were disconnected. The auroral displays over the mid-west of the USA were so bright that travellers on a trail in the Rocky mountains mistook the lights for dawn and began preparing breakfast, while gentlemen in New Orleans took advantage of the auroral light just after midnight to shoot birds flying from their roosts, 'innocently supposing it was daybreak'.

With the exception of the telegraph system, most Victorian-era technology was immune to the effects of space weather. But the Carrington event is useful in that it sets a realistic benchmark in terms of the severity of space weather. It leads to the obvious question: what would be the impact of such an event today? An obvious difference between Victorian and present-day technology is our exploitation of space. Satellite technology is ubiquitous, underpinning communications and navigation technologies and providing services for defence, meteorology, finance, and commerce to cite just a few examples. However, the space environment in which these satellites operate can be hostile, especially during period of inclement space weather. In a curious twist of fate, the radiation belts discovered by James Van Allen lie roughly the same distance from the Earth as the geostationary orbit (the distance at which satellites orbit the Earth once per day, meaning that satellites appear fixed relative to a location on the surface of the rotating Earth). During increased solar activity, the radiation belts can swell, engulfing satellites in geostationary orbit with an increased flux of high-energy sub-atomic particle. So-called 'killer electrons' moving at a fraction of the speed of light can damage the solar cells that provide a satellite's power, cause electric discharges that damage sensitive electronics, and deposit charge within the memory of semiconductor devices, generating phantom commands. Although far from the radiation belts, satellites in low Earth orbit are not safe from space weather effects. The deposition of increased amounts of energy in the Earth's upper atmosphere causes it to expand, increasing the atmospheric drag acting upon satellites and potentially perturbing their orbits, increasing the risk of premature re-entry and making collisions (due to incorrect positioning) more likely.

Increasing Need for Foresight

Our society's increased reliance on aviation for transport is another area where our vulnerability to space weather has increased. Aircrews and passengers spend significant periods above the densest (and most projective) portion of the lower atmosphere and risk exposure to increased radiation doses. Meanwhile, the ionosphere's refractive and transmissive effects on radio waves determines the viability of many communications links, but these are liable to change rapidly due solar activity. Both of these factors have an impact on the airline industry's increased exploitation of fuel- and cost-saving transpolar routes, where both radiation dosage and communication links are especially influenced by space weather conditions.

But perhaps the area where foresight is most required will be the impact of space weather on the electricity supply infrastructure. Indeed, one of the most significant societal risks posed by space weather is the impact of geomagnetically induced currents (GICs) on electricity distribution grids. The large magnetic field disturbances that occur during a geomagnetic storm (and caused by the impact of an ICME on the

Earth's magnetosphere) can induce electric fields in the solid Earth. Conducting networks (e.g. wires and pipelines) offer a low-resistance pathway for currents to flow across the surface of the Earth, giving rise to GICs in the network. Once flowing through a power network, GICs are unwanted quasi-direct currents, superimposed on the alternating currents within the grid, unbalancing and damaging critical transformers. However, after entering a conducting network via grounding points, the different pathways taken by GICs are influenced by the electrical properties of each network. As such, the study of GIC impact on national power grids incorporates aspects of geophysics, solar physics, solar-terrestrial physics, and power engineering.

There is much documented and anecdotal evidence of the effects of GICs on the power systems of the developed world. Possibly the most often cited example of a damaging impact is the collapse of the Hydro Quebec power system on 13 March 1989. A severe geomagnetic storm shut down the complete high-voltage system of Quebec in less than a minute, with significant knock-on economic cost and social disruption. More recent storms, for example, the October 2003 'Halloween' magnetic storm (which resulted in low-latitude auroral activity, including over the UK) are also known to have affected networks in Europe, North America, South Africa and elsewhere. A recent study has investigated the economic impact of a present-day repeat of the 'Carrington event': the cost in the USA alone is estimated at $1–2 trillion in the first year after the storm, with full recovery taking between four and ten years depending upon the level of damage to electricity supply infrastructure.

It is therefore unsurprising that governments, industry, and the space research communities are considering the likelihood and impact of space weather (Figure 6.4). Extreme space weather, of the kind that Carrington observed and that is likely to have a significant impact on society and the economy, probably occurs every few hundred years. Such high-impact, low-frequency events present unique challenges to policy makers. Comparable scenarios include pandemic diseases or, to cite a recent example, the massive air travel disruption caused that by the ash cloud associated with the 2010 eruption of the Eyjafjallajökull volcano in Iceland.

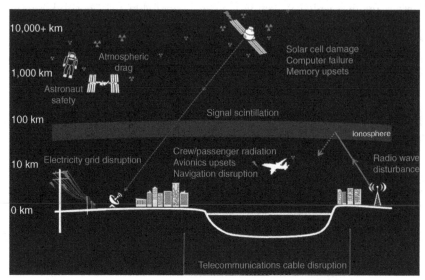

FIGURE 6.4 Space Weather Effects

In the spring of 2011, US President Obama met with UK Prime Minister Cameron and collaborative efforts between the USA and UK in the areas of space weather research and services were included in their discussions. The threat of disruption to the UK grid has been included in the current UK National Security Strategy and severe space weather has recently been added to the UK National Risk Register. Following a Parliamentary Select Committee investigation into scientific advice and evidence in emergencies, a Space Environment Impacts Expert Group has been set up to provide advice on space weather to the Chief Scientific Advisor in the Cabinet Office. In response, UK research councils are currently considering how best to support the basic research required to address the key outstanding science questions in this area and other UK agencies (e.g. the UK Meteorological Office) are considering the modelling required to deliver a space weather forecasting capability.

Meanwhile, learned societies and the insurance industry are undertaking research and presenting evidence-based advice to government and industry stakeholders.

One could argue that true foresight will come with the capability to accurately forecast space weather, in a fashion analogous to terrestrial meteorology. As in the terrestrial case, networks of measurements and computational models will hold the key, but the scarcity of space weather data is a challenge. For example, there are currently several hundred ground-based instruments and a few dozen satellites making measurements of the upper ionosphere and magnetosphere relevant to space weather. In contrast, fifteen thousand ground stations and over a thousand balloon flights daily are used to monitor terrestrial weather, despite the fact that the volume of the Earth's atmosphere is more than two hundred thousand times smaller than the terrestrial magnetosphere.

At present, the demand for decision-enabling space weather information is exceeding the research community's ability to provide it. However, an awareness of the space weather hazard and a new appreciation of the risks it poses to our society is emerging and stimulating research into forecasting techniques. Like other major natural hazards, foresight built upon evidence-based research will prove to be essential in developing operational forecasts and engineering solutions to mitigate the impact on our society and economy.

Further Reading

Cabinet Office (2012), National Risk Register of Civil Emergencies

Carrington, R.C. (1859) Description of a Singular Appearance seen in the Sun on September 1, 1859, *Monthly Notices of the Royal Astronomical Society* 20, 13–15

Forbes, K.F., and O.C. St Cyr (2008) 'Solar activity and economic fundamentals: Evidence from 12 geographically disparate power grids', *Space Weather* 6, S10003

Gaunt, C.T., and G. Coetzee (2007) 'Transformer failures in regions incorrectly considered to have low GIC-risk', in Proceedings of the IEEE Powertech Conference, July 2007, Lausanne, Switzerland

Green, J.L., S. Boardsen, S. Odenwald, J. Humble, and K.A. Pazamickas (2006) 'Eyewitness reports of the great auroral storm of 1859', *Advances in Space Research* 38, 145

H.M. Government. A Strong Britain in an Age of Uncertainty – National Security Strategy, 2010

Jim Wild

Lloyds (2010), Space Weather: Its impact on Earth and implications for business

Pulkkinen, A., S. Lindahl, A. Viljanen, and R. Pirjola (2005), 'Geomagnetic storm of 29–31 October 2003: Geomagnetically induced currents and their relation to problems in the Swedish high-voltage power transmission system', *Space Weather* 3, S08C03

Thomson, A.W.P., A.J. McKay, E. Clarke, and S.J. Reay (2005) 'Surface electric fields and geomagnetically induced currents in the Scottish power grid during the 30 October 2003 geomagnetic storm', *Space Weather* 3, S11002

Tsurutani, B.T., W.D. Gonzalez, G.S. Lakhina, and S. Alex (2003) 'The extreme magnetic storm of 1–2 September 1859', *Journal of Geophysical Research* 108, 1268

US National Research Council (2008), Severe space weather events – understanding societal and economic impacts. Committee on the Societal and Economic Impacts of Severe Space Weather Events: A Workshop, National Research Council, USA

Royal Academy of Engineering (2013), Extreme space weather: Impacts on engineered systems and infrastructure

7 How Individuals' Future Orientation Makes a Difference to Their Society: Self-control in a Four-decade Study of 1000 Children

TERRIE E. MOFFITT

Developing self-control over our behaviors, emotions, and thoughts is an essential human achievement. This achievement may be more essential now than ever before, because modern life seems to be increasing the value of self-control. Modern history has seen an extension of human longevity, requiring individuals to pay more strategic attention to their own health and wealth to avoid disability, dependency, and poverty in old age. Self-management of our own retirement savings demands of us foresight, planning, forbearance, and steadfastness. Today's seductive advertising technology calls for diligence and fortitude to resist the ubiquitous temptation to spend. Recent history has seen a marked increase in food availability and sedentary occupations, which obliges us to apply more self-discipline to keep fit and avoid obesity. Easy access to harmful addictive substances brings more need for willpower and moderation to avoid addiction. Greater ease of divorce requires more responsibility, patience, and resistance to desire, to prevent family dissolution. With mothers typically in the workforce, there is less time for parenting so parents strive to be more planful and conscientious in their approach to childrearing. Harsher laws, promoting imprisonment of young law-breakers, call for scrupulous impulse control, lest young people be caught up in the punitive criminal justice process. Greater urban population density increases the need for personal orderliness and constraint. As more and more citizens receive more and more years of education, intellectual achievement no longer wins the competition for

This research received support from the US NIA (AG032282; 2AG21178), the UK MRC (MR/K00381X), the New Zealand Health Research Council, and the Jacobs Foundation.

137

good jobs: employers now screen graduates for signs of conscientious-ness and perseverance. These remarkable historical shifts are enhancing the value of individual self-control in modern life, not just for wellbeing, but for survival. Unfortunately, not everyone develops good-enough self-control.

Building these valuable self-control skills begins in early childhood. The need to delay gratification, control impulses, and modulate emotional expression is the earliest and most ubiquitous demand that societies place on their children, and success at many life tasks depends critically on a child's mastery of such self-control. In this 2013 Darwin lecture, I describe a study of self-control in the lives of 1000 children. By age 10 years, many of them had mastered self-control but others were failing to achieve this skill. We followed them to the fourth decade of life and traced the consequences of their childhood self-control for their health, wealth, parenting, and criminal offending.

Interest in self-control unites all of the social and behavioral sciences. Self-control is an umbrella construct that bridges concepts and measure-ments from different disciplines (e.g. impulsivity, conscientiousness, self-regulation, delay of gratification, inattention-hyperactivity, executive function, willpower, inter-temporal choice). Neuroscientists study self-control as an executive function subserved by the brain's frontal cortex, and have uncovered brain structures and systems involved when research participants exert self-control inside the MRI scanner. Behavioral geneticists have shown that self-control is under both genetic and envir-onmental influences, and are now searching for genes associated with self-control. Psychologists have described how young children develop self-control skills, and traced population patterns of stability and change in self-control across the life course. Health researchers report that self-control predicts early mortality, psychiatric disorders and unhealthy behaviors such as over-eating, smoking, unsafe sex, drunk driving, and noncompliance with medical regimens. Sociologists find that low self-control predicts unemployment, and criminologists name self-control as a central causal variable in crime theory, providing evidence that low self-control characterizes law-breakers.

Economists are now drawing attention to individual differences in self-control as a key consideration for policy-makers who seek to enhance the physical and financial health of the population, promote child welfare, and reduce the crime rate. The current emphasis on self-control skills of conscientiousness, self-discipline, and perseverance arose from the empirical observation that preschool programs that targeted poor children 50 years ago, although failing to achieve their stated goal of lasting improvement in children's IQ scores, somehow have produced surprising by-product reductions in teen pregnancy, school dropout, delinquency, and work absenteeism. To the extent that self-control influences outcomes as disparate as health, wealth, parenting, and crime, enhancing it could have broad benefits. Given that self-control is malleable, it could be a prevention target, and the key policy question becomes when to intervene to achieve the best cost–benefit ratio, in childhood or in adolescence? However, regardless of its malleability, if low self-control is influential, policy-makers might exploit this by enacting so-called opt-out schemes that tempt people to eat healthy food, save money, and obey laws, by making these the easy default options that require no effortful decision-making or self-control. If citizens were obliged to take action to opt out of default health-enhancing programs or payroll-deduction retirement savings schemes, individuals with the lowest self-control should tend to take the easy option and stay in programs, because opting out requires of them unappealing effort and planning. Similarly, the idea behind the crime-reduction policy of 'target hardening' is to discourage would-be offenders by making law-breaking require more effortful planning (e.g. sophisticated anti-theft devices in cars require more advance planning and diligence of car thieves).

The Dunedin Developmemt Study

In the context of this timely, ubiquitous, and intense policy interest in self-control, we report findings from the Dunedin Multidisciplinary Health and Development Study, a longitudinal investigation of health and behavior in a birth cohort. The 1037 Study members were all individuals born between April 1972 and March 1973 in Dunedin, New Zealand, who were eligible for the longitudinal study based on residence in the province when they were age 3 years and who participated in the first follow-up assessment then (91% of eligible births; 52% male). The cohort represents the full range of socioeconomic status in the general population of New Zealand's South Island and is primarily white. Assessments have been carried out at ages 3, 5, 7, 9, 11, 13, 15, 18, 21, 26, 32, and 38 years. In 2011–2012, 95 percent of the surviving 1007 Study members took part in assessment. At each assessment wave, Study members are brought to the Dunedin research unit for a full day of interviews and examinations. These data are supplemented by searches of official administrative records and by questionnaires that are mailed, as developmentally appropriate, to parents, teachers, and peers.[1]

The design of the Dunedin Study is observational and correlational; this is in contrast to experimental behavioral-economics studies that ascertain the association between performance on laboratory self-control tasks (such as delay of gratification, discounting, and inter-temporal choice tasks) and behavioral proxy measures of wealth, health, and crime. Such laboratory experiments yield compelling information about self-control, although economists have debated whether behavior in the lab faithfully represents real-world behavior. The naturalistic Dunedin Study complements experimental laboratory research on self-control by providing badly needed information about how well children's self-control, as it is distributed in the population, is actually predicting real-world outcomes after children reach adulthood.

The Dunedin Study's birth-cohort members with low self-control and poor outcomes have not dropped out of the study. This enabled us to study the full range of self-control and to estimate effect sizes of associations for the general population, information that is requisite for informed policy-making. The Dunedin Study's design allowed us to address multiple policy-relevant hypotheses. First, we tested whether children's self-control predicted later health, wealth, parenting, and crime similarly at all points along the self-control gradient, from lowest to highest self-control. If self-control's effects follow a gradient, then interventions that achieve even small improvements in self-control for individuals could shift the entire distribution of outcomes in a salutary direction and yield large improvements in health, wealth, child-welfare, and crime-reduction for a nation. Second, although this study did not include an intervention, some Dunedin Study members moved up in the

---

[1] Learn more about the Dunedin Study at http://dunedinstudy.otago.ac.nz/.

self-control rank over the years of the study, and we were able to test the hypothesis that improving self-control is associated with better adult life outcomes. Third, because we assessed whether study members as adolescents smoked tobacco, left secondary school early, or became teen parents, we were able to test the hypothesis that children with low self-control make these mistakes as teenagers that close doors of opportunity and ensnare them in lifestyles harmful to their health, wealth, parenting skills, and the public's safety. If self-control's influence is mediated through adolescents' mistakes, adolescence could be an ideal window for intervention policy. Fourth, because the Dunedin Study assessed self-control as early as age 3 years, we were able to test the hypothesis that individual differences in preschoolers' self-control predict outcomes in adulthood. If so, early childhood would also be a good intervention window. Fifth, to address concerns that children with the highest levels of self-control must be constricted, rigid, and incapable of spontaneity or happiness, we tested whether our study members' childhood self-control predicted their creative accomplishments and satisfaction with their lives as adults.

We assessed children's self-control during their first decade of life. Behavior ratings tapped items such as: emotional lability, flying off the handle, low frustration tolerance, lacking persistence, short attention span, distractibility, shifting from activity to activity, restlessness, being over-active, poor impulse control, acting before thinking, difficulty waiting, difficulty in turn-taking. Reports of these kinds of behaviors were provided by researcher-observers, teachers, parents, and the children themselves, across ages 3, 5, 7, 9, and 11 years. Because these many ratings were strongly correlated, we combined them into a single highly reliable composite self-control measure. All children lack self-control now and then, but this composite measure insured that low scorers had shown poor self-control in different situations (home, research lab, and school) and across the years from age 3 to 11 years. Mean levels of self-control were significantly higher among girls than boys of course, but the health, wealth, child-welfare, and public-safety implications of childhood self-control were equally evident for both sexes. We therefore combined the sexes in all subsequent analyses (but controlled statistically for sex).

Terrie E. Moffitt

Correlates of Self-control

It is well known that high family social class and good intelligence
influence children's adult life success, but social class and IQ have proven
resistant to intervention. Dunedin children with greater self-control were
significantly more likely to have been brought up in socioeconomically
advantaged families. Children with greater self-control also had significantly higher tested IQs on average. These two findings raised the
question of whether low self-control has any influence on outcomes
apart from low social-class origins or low intelligence. We thus introduced statistical controls to all analyses described here, to test whether
childhood self-control predicted adults' health, wealth, parenting, and
crime independently of their social-class origins and IQ.

We examined adult health problems because these are known early-
warning signs for costly age-related diseases and premature mortality.
When the children reached their thirties, we assessed their cardiovascular,
respiratory, dental, and sexual health, as well as their inflammatory status
by carrying out physical examinations and laboratory tests to assess meta-
bolic abnormalities (including overweight, hypertension, and cholesterol),
lung airflow limitation, periodontal disease, sexually transmitted infection,
and blood C-reactive protein levels. We summed these five clinical measures into a simple physical health index for each Study member; 43 percent
of Study members had none of the clinical biomarkers, but 20 percent had
two or more biomarkers. Childhood self-control significantly predicted the
number of these adult health problems (Figure 7.1).

We also conducted clinical interviews with the Study members in their
thirties to assess substance dependence (tobacco, alcohol, and cannabis
dependence, as well as dependence on other street and prescription
drugs). As adults, children with poor self-control had significantly ele-
vated risk of substance-dependence. This longitudinal link between self-
control and substance-dependence was verified by people that Study
members had nominated as informants who knew them well. As adults,
more children with poor self-control were rated by their informants as
having alcohol and drug problems. Both of the aforementioned health
predictions followed the gradient, and both remained significant after
statistically accounting for social-class origins and IQ.

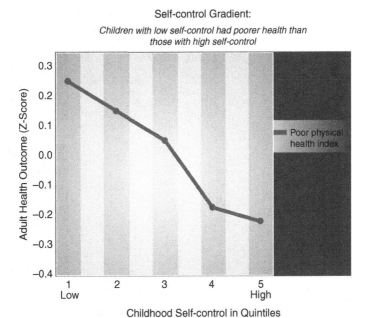

Self-control Gradient:

Children with low self-control had poorer health than those with high self-control

FIGURE 7.1 Adult Health

We examined wealth outcomes such as low income, poor saving habits, credit problems, and social welfare benefit dependency, because these are early warning signs for late-life poverty and financial dependence. Childhood self-control foreshadowed the Study members' financial situations when they reached their thirties. Although the Study members' social-class of origin and IQ were strong predictors of their adult socioeconomic status and income as expected, poor self-control offered significant incremental validity in predicting the socioeconomic position of their occupation and the income they earned (Figure 7.2). In their thirties, children with poor self-control were significantly less financially planful: Compared to their cohort peers they were less likely to save and they had acquired fewer financial building blocks for the future (such as home ownership, investment funds, or retirement plans). Children with poor self-control were also struggling financially in adulthood: They reported to us more money-management difficulties and more credit problems.

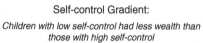

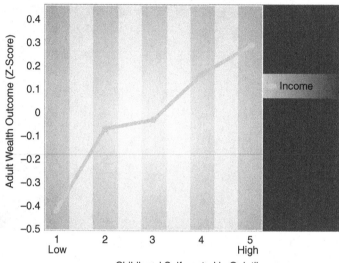

FIGURE 7.2 Adult Wealth

This longitudinal link between self-control and self-reported financial problems was verified by informants who knew them well: As adults, children with poor self-control were rated by their informants as poor money managers. Official administrative records also confirmed that childhood self-control predicts adult financial outcomes. For example, self-control ratings in childhood foretold our Study members' official credit ratings in adulthood, ratings which have real consequences for their ability to borrow business capital or obtain a home mortgage. A match of Study members to the VEDA Credit System for Australia and New Zealand revealed that the lowest self-control children were most likely to be rated as undesirable credit risks as adults. With the assistance of the New Zealand Ministry of Social Development we further examined costs to government in the form of social welfare benefit dependency, by matching Study members to administrative records of monthly social welfare payments. More than half the cohort had at some time received such benefit payments. Self-control only weakly predicted which cohort members had ever received a benefit, indicating that many people need a helping hand and can benefit from a safety net, particularly in today's weak economy. However if they did receive a benefit, those Study members with the poorest self-control were likely to stay on benefits for a longer period of time. Benefit recipients in the cohort's highest quintile in childhood self-control had used benefits on average under 18 months, whereas recipients in the lowest self-control quintile had used benefits on average for more than 6 years. Again, all of these findings about the Study members' financial lives followed the gradient, and all held after statistical controls for social class background and IQ.

The first Study member became a parent in 1988 at age 15, and by 2012 at the end of their thirties, three quarters of Study members had become parents. As each Study-member parents' first child reaches age 3 years, the research team visits them at home to record videotapes of parent–child interaction during a standardized set of activities. Videos were rated by family psychologists blind to all other information about the Study, for aspects of parenting that include warmth and affection, providing stimulating support for toddler development, and sensitivity to their child's needs. On an overall combined rating of parenting quality, Study members who as children had poor self-control had grown up to be the least-skilled parents of their own children. Childhood self-control also predicted whether or not these Study members' offspring were being reared in a two-parent versus a one-parent household (e.g. the Study member was an absent father or solo mother). These parenting findings too followed the gradient, and they held after accounting for social-class and IQ.

We also examined convictions for crime, because crime control poses major costs to government. We obtained records of Study members' court convictions at all courts in New Zealand and Australia by searching the central computer systems of the New Zealand police; one quarter of the Study members had been convicted of a crime by their thirties (matching rates in other developed nations). Children with poor self-control were more likely to have been convicted of a criminal offense, even after accounting for social-class origins and IQ (Figure 7.3).

Among the 5 percent of the cohort who had spent time incarcerated, more than 80 percent came from the cohort's two lowest childhood self-control quintiles.

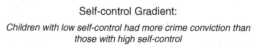

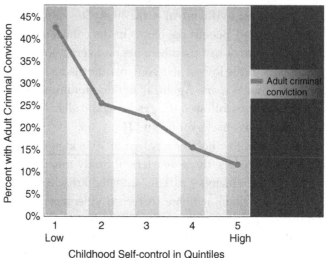

FIGURE 7.3 Adult Crime

The Self-control Gradient

We observed a *self-control gradient* in which boys and girls with less self-control had worse health, less wealth, less-skilled parenting, and more crime as adults than those with more self-control, at every level of the distribution of self-control (Figures 7.1–7.3). To further document that self-control relates to outcomes all along its gradient, we tested whether self-control effects operate throughout the population distribution versus are confined to the least self-controlled children. We repeated analyses after removing children in the least self-controlled quintile, and continued to observe significant linear associations. The self-control gradient was even apparent when we removed children in the least and most self-controlled quintiles. To ask if the societal challenge of self-control could be ameliorated by treating childhood attention deficit hyperactivity disorder (ADHD; a childhood psychiatric disorder of impaired impulse-control) we removed the 61 Study members who were diagnosed with ADHD, and repeated the analyses. The gradient associations remained unaltered. These results, coupled with the robust effects of self-control after controlling for variation in sex, social-class origins, and IQ, suggest that enhancing self-control skills can benefit even intelligent children from well-to-do homes, as well as children who score above average on self-control.

Can interventions with adolescents break the connections from childhood self-control to adult health, wealth, parenting, and crime? Data collected at ages 13, 15, 18, and 21 years showed children with poor self-control were more likely to make mistakes as adolescents, resulting in 'snares' that ensnared them in harmful lifestyles. For example, more low-self-control children began smoking by age 15, left secondary school early with no educational qualifications, and became unplanned teenaged parents (Figure 7.4). (Our choice of snares was not exhaustive, but we elected to study those which are already high-priority targets of adolescent education policy.) The lower their self-control the more of these snares they encountered. In turn, the more snares they encountered, the more likely they were, as adults, to have poor health, less wealth, unskilled parenting, and a criminal-conviction record. We tested whether adolescent snares explained the long-term predictive power of self-control in two ways: First, using statistical controls, we partialled out the portion of the association between childhood self-control and each adult outcome that was accounted for by adolescent snares. The snares attenuated the effect of self-control on outcomes by approximately 50 percent. However, a direct effect of self-control remained statistically significant for nearly every outcome measure. Second, we tested the association between childhood self-control and the adult outcomes among adolescents who did not encounter any snares, a so-called 'utopian group' of non-smoking, non-teen parent, secondary-school graduates (20). Compared to the full cohort, this utopian sub-set of Dunedin Study members had fewer health problems, more wealth, and less crime (Figures 7.5, 7.6, and 7.7), illustrating that preventing adolescent mistakes could enhance adult outcomes for children at every level of childhood self-control.

This comparison shows what a successful intervention might accomplish. However, prediction from childhood self-control to the adult measures remained significant along a gradient, even among this lucky group who escaped adolescence without becoming ensnared by such mistakes.

Getting Trapped by Adolescent Mistakes:

● Children with low self-control were more likely to make mistakes as teen…

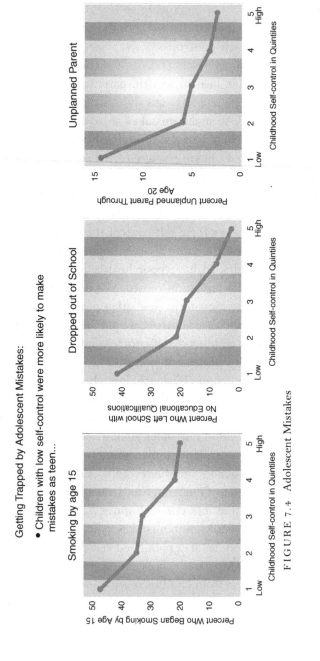

FIGURE 7.4 Adolescent Mistakes

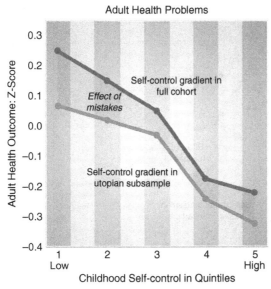

FIGURE 7.5   Adolescent Mistakes and Adult Health

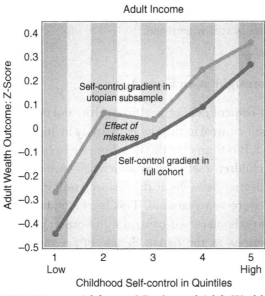

FIGURE 7.6   Adolescent Mistakes and Adult Wealth

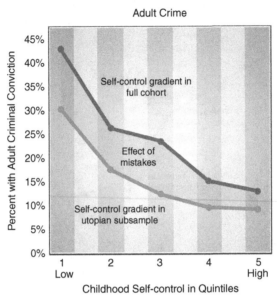

FIGURE 7.7 Adolescent Mistakes and Adult Crime

Helping teens avoid snares could improve population health, wealth, child-welfare, and public safety somewhat, but building self-control skills before the teen years is still warranted.

Social Foresight at an Early Age

How early can self-control predict health, wealth, and crime? Our composite measure of self-control in the Dunedin Study included assessments from age 3 to age 11. But to answer this question, we isolated our research staff ratings of the children's self-control which had been observed during 90-minute data-collection sessions at the research unit in the mid-1970s, when the children were 3–5 years old. This brief standardized observational measure of preschoolers' self-control significantly predicted health, wealth, and convictions in the fourth decade of life, albeit with more modest effect sizes.

But are they happy? Some audiences who hear of our findings have expressed concern that children with the highest levels of self-control must be constricted, rigid, and incapable of spontaneity or happiness. Prompted by this concern, we examined indicators of our study members' life satisfaction in the fourth decade of life. Fortunately, most cohort members said they are satisfied with their lives: 70 percent were somewhat or very satisfied. However, the most satisfied of all were those who began life with high self-control (over 90 percent felt satisfied).

Policy-making requires evidence that isolates self-control as the active ingredient affecting health, wealth, and crime, as opposed to other influences on children's futures, such as their intelligence or social-class origins. In the Dunedin Study, statistical controls revealed that self-control had its own associations with outcomes, apart from childhood social-class and IQ. However, each Dunedin Study member grew up in a different family, and their families varied widely on many other features that influence children's life outcomes. We also exploited another longitudinal study, a British birth cohort of siblings, to ask whether the sibling in each pair who had lower self-control subsequently developed worse outcomes, despite both siblings having the same neighborhood, home and family. This compelling quasi-experimental research design can isolate the influence of self-control by tracking and comparing siblings to disentangle the individual child's self-control from all other features on which families differ (and which the siblings share while growing up).

The British E-risk Twins Study

To apply this design, we turned to a second sample, the Environmental-Risk Longitudinal Twin Study (E-risk), where we have been tracking a birth cohort of British twins since their birth in 1994–95, with 96 percent retention. When the E-Risk twins were 5 years old, our research staff rated each child on the same observational measure of self-control originally used with Dunedin children as preschoolers. Although the E-risk children had been followed only up to age 12 years when we performed this analysis, their self-control already forecasted many of the adult outcomes we saw in the Dunedin Study. We applied sibling fixed-effects models to 509 same-sex dizygotic (fraternal) twin pairs, because they are no more alike than ordinary siblings (with the added advantages of being the same age and sex). Models showed that the

151

5-year-old sibling with relatively poorer self-control was significantly more likely as a 12-year-old to begin smoking (a precursor of adult ill-health), perform poorly in school (a precursor of adult wealth), and engage in antisocial conduct problems (a precursor of adult crime), and these findings remained significant even after controlling for sibling differences in IQ.

An especially provocative finding from the British E-Risk study attests to the societal cost of low self-control. The twin with lower self-control was rated by teachers as requiring more of their effort in the classroom. E-Risk twins' teachers provided answers to questions about teaching effort, including 'Compared to classmates, how often must you act to curb disruptive behaviour by this child?,' 'Compared to classmates, how often does this child require one-on-one attention from you?,' and 'Compared to classmates, how often does this child's behavior make you feel frustrated?' Results showed that children with low self-control take teachers' energy away from teaching other pupils. Children lacking in self-control may even contribute to teachers' job dissatisfaction and loss of teachers to the profession, at potential great cost to the education system.

We have learned from this research that differences between individuals in self-control are present in early childhood and can predict multiple indicators of health, wealth, and crime, across three decades of life, in both sexes. Furthermore, it was possible to disentangle the effects of children's self-control from effects of variation in the children's intelligence and the social-class and home lives of their families, thereby singling out self-control as a clear target for intervention policy. Joining earlier longitudinal follow-up studies (5, 7, 21), our findings imply that innovative policies that put self-control center stage might reduce the panoply of costs that now heavily burden citizens and governments.

Differences between children in self-control predicted their adult outcomes approximately as well as low intelligence and low social-class origins, which are known to be extremely difficult to improve through intervention. Effects were marked at the extremes of the self-control gradient. For example, by adulthood, the highest and lowest fifths of the population on measured childhood self-control had respective rates of multiple health problems of 11 percent versus 27 percent; rates of poly-substance dependence of 3 percent versus 10 percent; rates of annual income under NZ$20,000 of 10 percent versus 32 percent; rates of off-spring reared in single-parent households of 26 percent versus 58 percent; and crime-conviction rates of 13 percent versus 43 percent. This coincidence of low self-control with poor outcomes bolsters the rationale for 'opt-out' programs, by demonstrating that the segment of the adult population that is most inclined to avoid the effortful planning necessary to opt out of default programs (i.e. individuals with the lowest self-control) is the same segment of the adult population that accounts for excess costs to society in health care, financial insolvency, social welfare dependency, and crime. Opt-out programs intended to exploit the laziness in all of us should work best for the least conscientious among us, significantly magnifying their benefits for the whole population.

Fostering More Self-control

With respect to timing of programs to enhance self-control, our findings were consistent with a 'one-two punch,' scheduling interventions during both early childhood and adolescence. On the one hand, low self-control's capacity to predict health, wealth, and crime outcomes from childhood to adulthood was, in part, a function of mistakes our research participants made in the interim adolescent period. Low-self-control adolescents made mistakes such as starting smoking, leaving high school, and having an unplanned baby that could ensnare them in lifestyles with lasting ill effects. Thus, interventions in adolescence that prevent or ameliorate the consequences of teenagers' mistakes might go far to improve the health, wealth, child welfare, and public safety of the population. On the other hand, that childhood self-control predicts these adolescent mistakes implies early childhood intervention could prevent them ever occurring in the first place. Moreover, even among teenagers who managed to finish high school as non-smokers and non-parents, the level of personal self-control they had achieved as children still explained variation in their health, finances, and crime when they reached their thirties. Early childhood intervention that enhances self-control is likely to bring a greater return on investment than harm-reduction programs targeting adolescents alone.

With respect to the scope of programs addressing self-control, our findings raise the question of whether early intervention to enhance self-control should take a targeted approach to a few children extremely lacking in self-control versus a universal approach to all children. Health, wealth, and crime outcomes followed a gradient across the full distribution of self-control in the population. If correct, the observed gradient implies room for better outcomes even among the segment of the population whose childhood self-control skills were somewhat above-average, and even among intelligent children and children from well-to-do homes. Universal interventions that benefit everyone often avoid stigmatizing anyone and also attract widespread citizen support. Testing this gradient in other population representative samples is a research priority. It has been shown that self-control can change. In fact, in the Dunedin Study, across the first four decades of life measured self-control was only about half as stable as the IQ. Programs to enhance children's self-control have been developed and positively evaluated, and the challenge remains to improve these programs and scale them up for universal dissemination. Understanding the key skills of self-control and how to best enhance them with a good cost–benefit ratio is a research priority.

This year's theme for the Darwin Lectures is foresight. The aging of the human population allows demographic predictions that suggest useful foresight: the value that society will place on every child will increase this century, along with an increase in societal expectations that each child must contribute his or her very best to their nation's health, wealth, child welfare, and public safety. The population is trending toward fewer children and more elderly people, indicating that each young worker will be obliged to support more old people. At the same time, the population is trending toward longer life expectancy, indicating that today's children must prepare strategically while yet young, to secure wellbeing in their own protracted old age. That many Dunedin Study members with low self-control had unplanned babies who are now growing up in low-income single-parent households lacking in warm, skilled parenting reveals that one generation's low self-control disadvantages the coming generation. Two cohorts born in different countries and different eras support the inference that individuals' self-control, beginning in childhood, is a key ingredient in health, wealth, child welfare, and public safety, and a sensible policy target.

Further Reading

Bogg, T. and B.W. Roberts (2004) 'Conscientiousness and health behaviours: A meta-analysis', *Psychology Bulletin* 130, 887–919

Bouchard, T.J. (2004) 'Genetic influence on human psychological traits', *Current Directions in Psychological Science* 13, 148–151

Caspi, A., T.E. Moffitt, D.L. Newman, and P.A. Silva (1996) 'Behavioural observations at age 3 years predict adult psychiatric disorders: Longitudinal evidence from a birth cohort', *Archives of General Psychiatry* 53,1033–1039

Caspi, A., *et al.* (1994) 'Are some people crime-prone: Replications of the personality–crime relationship across countries, genders, races, and methods', *Criminology* 32, 163–195

Caspi, A., B.R.E. Wright, T.E. Moffitt, and P.A. Silva (1998) 'Early failure in the labor market: Childhood and adolescent predictors of unemployment in the transition to adulthood', *American Sociology Review* 63, 424–451

Diamond, A. (2011) 'Interventions shown to aid executive function development in children 4 to 12 years old', *Science* 333, 959–964

Ebstein, R.B. (2006) 'The molecular genetic architecture of human personality', *Molecular Psychiatry* 11, 427–445

Eslinger, P.J., C. Flaherty-Craig, and A.L. Benton (2004) 'Developmental outcomes after early prefrontal cortex damage', *Brain and Cognition* 55, 84–103

Gottfredson, M. and T. Hirschi (1990) *A General Theory of Crime.* Stanford, CA: Stanford University Press

Greenberg, M.T. (2006) 'Promoting resilience in children and youth: Preventive interventions and their interface with neuroscience', *Annals of the New York Academy of Sciences* 1094, 139–150

Hare, T.A., C.F. Camerer, and A. Rangel (2009) 'Self-control in decision-making involves modulation of the vmPFC valuation system', *Science* 324, 646–648

Heckman, J. (2007) 'The economics, technology, and neuroscience of human capability formation', *Proceedings of the National Academy of Science* 104, 13250–13255

Heckman, J.J., J. Stixrud, and S. Urzua (2006) 'The effects of cognitive and noncognitive abilities on labor market outcomes and social behaviour', *Journal of Labor Economics* 24, 411–482

Heckman, J.J. (2006) 'Skill formation and the economics of investing in disadvantaged children', *Science* 312, 1900–1902

Jackson, J.J., *et al.* (2009) 'Not all conscientiousness scales change alike: A multimethod, multisample study of age differences in the facets of conscientiousness', *Journal of Personality and Social Psychology* 96, 446–459

Kern, M. and H. Friedman (2008) 'Do conscientious individuals live longer? A quantitative review', *Health Psychology* 27, 505–512

Knudsen, E.I., J.J. Heckman, J.L. Cameron, and J.P. Shonkoff (2006) 'Economic, neurobiological, and behavioural perspectives on building America's future workforce', *Proceedings of the National Academy of Science* 103, 10155–10162

Kochanska, G., K.C. Coy, and K.T. Murray (2001) 'The development of self-regulation in the first four years of life', *Child Development* 72, 1091–1111

Layard, R. and J. Dunn (2009) *A Good Childhood: Searching for Values in a Competitive Age*. London: Penguin

Mischel, W., Y. Shoda, and M. Rodriguez (1989) 'Delay of gratification in children', *Science* 244, 933–938

Moffitt, T.E., L. Arseneault, D. Belsky, N. Dickson, R.J. Hancox, H.L. Harrington, R. Houts, R. Poulton, B. Roberts, S. Ross, M. Sears, M. Thomson, and A. Caspi (2011) 'A gradient of childhood self-control predicts health, wealth, and public safety', *Proceedings of the National Academy of Science* 108, 2693–2698.

National Scientific Council on the Developing Child (2007) *The Science of Early Childhood Development*, http://www.developingchild.net

Oeppen, J. and J.W. Vaupel (2002) 'Broken limits to life expectancy', *Science* 296, 1029–1031

Piquero, A.R., W.G. Jennings, and D.P. Farrington (2010) 'On the malleability of self-control: Theoretical and policy implications regarding a general theory of crime', *Justice Quarterly* 27, 803–834

Roberts, B.W., K. Walton, and W. Viechtbauer (2006) 'Patterns of mean-level change in personality traits across the life course: A meta-analysis of longitudinal studies', *Psychological Bulletin* 132, 1–25

Stuss, D.T. and D.F. Benson (1986) *The Frontal Lobes.* New York: Raven Press

Thaler, R.H. and C.R. Sunstein (2008) *Nudge: Improving Decisions about Health, Wealth, and Happiness.* New York: Penguin Group

White, J.L., *et al.* (1994) 'Measuring impulsivity and examining its relationship to delinquency', *Journal of Abnormal Psychology* 103, 192–205

8 Foresight in Ancient Mesopotamia

FRANCESCA ROCHBERG

Introduction

Assyria's place among ancient states of the Near East has been secured for modern historians by its imperial achievement. From the eighth to the end of the seventh centuries BC, Assyria organized under one rule an area encompassing southwestern Iran to southeastern Turkey (Figure 8.1). It engulfed the Levant, Egypt, and Babylonia, gaining access to the Persian Gulf, the Mediterranean Sea, and the rich trade routes through the Zagros Mountains east to Afghanistan and north into Urartu, or present-day Armenia. As the first and architect of all later empires, Assyria is known principally from the many monuments to its own sovereignty and power that survive in the form of exquisite inscribed stone bas-relief representations of the exploits of war, conquest, and cruelty to its enemies (Figure 8.2). These images, coupled with vivid accounts in the Bible, where Assyria was reviled by the prophet Isaiah, give us something of the face Assyria presented to the outside world. A more behind-the-scenes look at the internal workings of this first of all empires, however, takes us directly to the subject at hand, to the systematic institutional implementation of foresight for the management and maximization of the future power and security of the state of Assyria.

A passage from a seventh-century BC diviner's guide recovered from the Library of Assurbanipal in the Assyrian capital of Nineveh contains the following question: 'When you know the sign and they ask you to save the city, the king, and his population from the hands of the enemy, from pestilence, and famine, what will you say? They will say to you, how will you make (the evil consequences of omens) pass by?'

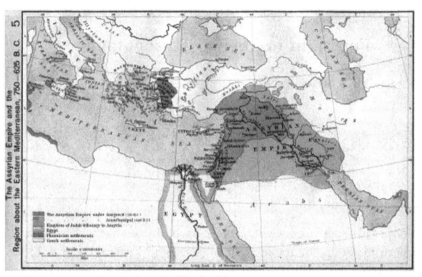

FIGURE 8.1 Assyrian Empire Map

FIGURE 8.2 Lachish Relief Reign of Sennacherib

The diviners of the royal court, well-trained *literati*, expert in the identification and interpretation of ominous signs, forecast the future on behalf of, as the text says, 'the city, the king, and his population.' According to letters from these diviners preserved from the reigns of the last two great Assyrian monarchs, Esarhaddon and Assurbanipal, important decisions concerning military, cultic, and personal matters were gauged in accordance with forecasts from ominous signs.

Ever since the mid-nineteenth century decipherment of the cuneiform script, the wedge-writing for which Sumerians, Babylonians, Assyrians, and the Achaemenid Persians are now famed, modern Assyriologists have worked to reconstruct the *history* of the Assyrian Empire. Our modern reconstruction focuses on its formation, exploitation of conquered territories, and final fall. This history has been reconstructed from excavated artifacts and images testifying to the geographical extent of Assyrian influence, and from data extracted from extant Assyrian accounts of wars, rebellions, and conquest. But the *story* of the Assyrian Empire, the one that the Assyrians told themselves, their descendants, and the gods, is one of quite different focus and emplotment. This internal version, or master narrative, is the story that underpins the annalistic inscriptions of Assyrian kings. It gives the reportage of campaigns, sieges, and victories over rebels a framework, revealing the importance of foresight as a central strategy for the conduct of imperial business. That narrative, not unlike what we see in the biblical histories, featured the gods as historical actors and focused on the intimate relationship between the king and his gods.

In the Assyrian understanding of both the historical past and the imagined future of the empire, the gods figured as prominently as the king's own royal ancestors and descendants. Inscriptions telling of the achievements of the Assyrian kings are as though reports directed to these gods, proving the dedication of the king in carrying out the divine will for him to rule, to expand his lands and amass wealth with which to adorn temples and divine statues. The means used by the crown to decide if and when to take action were divinatory techniques designed to gather knowledge of the future by communication with the gods. Modern scholars have mined Assyrian royal inscriptions for historical information, but the purpose of those inscriptions, and of many other sources that shed light on Assyrian political and military action, was not simply to render a historical or political account of events, but to legitimate the imperial agenda and project its security into the future by claims to divine approval.

Foresight: Ancient and Modern

Today the institutional variety of foresight is associated with various collectives, corporations, or polities, entities that need to make informed decisions about the future, specifically about how they might see their own success maximized in the future. Scenarios and narratives, visions and alternatives are part of the terminology of institutional foresight. As we try to understand the perspective from inside the Assyrian royal court, itself a complex collective entity spearheaded by the reigning monarch, these concepts are not entirely unrelated. Of course the premise on which foresight was institutionalized in Assyria stemmed from a worldview and a metaphysics that do not translate into those of Western modernity. The impulse to project an image of the future into which it could see its own growth and success, however, is certainly kindred. And for this to happen, the Assyrian king needed a large complement of experts in techniques of divination.

Celestial diviners and haruspices, who inspected the liver of a sacrificed sheep, each made forecasts of political and economic events on behalf of the king. These elite scribes in the employ of the Assyrian kings made reports in writing to the king, who then took the results of their liver inspection or celestial observation into account before undertaking many activities. It was not enough to know the signs and make forecasts from them, or even to predict, in the case of celestial signs, the very phenomena whose appearances were ominous. It was also necessary, as indicated in the passage just quoted, for the scribal specialist to, as the text says, 'make (the evil) pass by.' Foresight in ancient Assyria, therefore, not only involved forecasting and prediction, but also forestalling by apotropaic magic; that is, magic that wards off evil.

Then, as now, foresight is not simply looking forward, but anticipating future change and acting on that vision. Fundamental and essential to the system within which foresight was institutionalized by the Assyrian state was the idea that certain divinities communicated their knowledge of the future, indeed their will concerning the future, by producing the phenomena of the world, both on earth and in the sky, to function as indicators of future events. This all-important feature of foresight in cuneiform culture is the distant cousin to our own English usage meaning providence, or divine foresight. Contrary to Western assumptions about the inevitability of divine providence, sometimes associated with the natural order of the world itself, ancient Near Eastern gods were amenable to prayer and ritual appeasement, as evidenced in the many extant prayers and apotropaic rituals against evil portents from such things as the appearance of animals, snakes, malformed births, and lunar eclipses. Our diviner's handbook instructs that

> When you look up a sign (in the omen collections) be it one in the sky or one on earth and if that sign's evil portent is confirmed(?) ... in reference to an enemy or to a disease or to a famine, check the date of that sign and should no sign have occurred to counteract that sign, should no annulment have taken place, one cannot make it pass by, its evil cannot be removed. It will happen.

It is clear from this passage that the apotropaic effects of magic were thought to be efficacious in counteracting signs and removing the evil they portended. If no such steps were taken, the sign's portended consequences were then considered inevitable. This, however, is no fatalism. The Assyrians lived in a world of future contingencies, not of immutable determinism. Their gods knew the future, but the nature of divine foreknowledge was of a different order from that of the later Christian God, about whose foreknowledge and predetermination of all things mediaeval theologians and scholastics attempted to reconcile with the question of free will.

Foresight and fatalism do not sit well together, and the ancient Assyrian diviners were not hampered by notions of the fixity of nature or the chain of physical causality. Nonetheless, an omen functioned as an indicator of future events not as a one-time or random occurrence, put there by the gods *ad hoc* to signal particular events in the sense of this battle, or that disease. As conditional statements in the form 'If P then Q', phenomena were indicators of events in a general, even abstract, and regularly recurring way. Eclipses of a certain description portended plague, or the death of a monarch; the appearance of a star inside the halo of the moon signaled that the king and his troops would be surrounded and besieged, and so on, covering thousands of phenomena. Celestial phenomena and their portended events did not, however, stand in relation to one another as cause to effect. Rather, they stood as meaningful correlations, often construed by the use of analogy, as in the example of the star in the lunar halo indicating that the kings' army will be under siege. Ancient Mesopotamian omens reflect neither *post hoc ergo propter hoc* fallacious reasoning, nor a system of mechanical causality. They operated on the basis of association and analogy, or by other kinds of connections inscrutable to our way of thinking.

The foregoing description of omens as correlations or associations is very much a modern translation of how Assyro-Babylonian omens functioned to indicate the future. To the *literati* trained in reading and interpreting cuneiform texts, however, omens were a form of writing. Physical phenomena, either the marks and contours of the liver, or the various phenomena of celestial bodies, were, according to the scribes' own description, *written* messages concerning the future. The gods wrote on the liver – called the 'tablet of the gods' – and they wrote on the sky, where celestial phenomena were called 'heavenly writing.' The diviner-scribes, therefore, read the gods' heavens and the liver in the same manner as they read cuneiform tablets.

The divine writing is a compelling metaphor, if indeed it was a metaphor, for the intelligibility of the world both in heaven and on earth. The comprehensible nature of the world as manifested in a divine orthography was just as basic to the Assyrian and Babylonian conception of the world as the later notion of the Book of Nature became for Western natural philosophy and science. What is important here, however, is not just that physical phenomena were comprehensible, but that their value for gauging future prospects was based upon a premise of regular recurrence. Regular recurrence was not understood to manifest a system of causes, but of regular and recurring associations or meanings. In this way, the future was bound up with the past by the logic of its own internal semantics and orthography. By association, P indicated Q in the past, present, and future. Whenever P was observed, therefore, Q was also expected to occur.

The connection between the past and the future is undoubtedly a critical node in the logic of foresight. According to Alfred Marcus, who has worked on the principles, models, and methods of foresight in business management, organizations rely on master narratives to 'structure people's understanding of the past and the future.' What narratives, or stories, do best is lend coherence to what otherwise may be construed as random occurrences, or events whose connection is sometimes difficult to see. The future relates to the past by the logic of the narrative. With respect to the future's dependence upon the past, as Marcus put it, 'understanding what has occurred provides an impetus for what to do next. It supplies reasoning for affirming or rejecting old patterns of behavior and starting new projects that put organizations on novel trajectories. The past is a critical and essential springboard for thinking about the future.'

Such is the nature of Mesopotamian omen divination, which, in turn, informed the internal narrative of the Assyrian Empire. The signs signaled future events because the Assyrians were committed to the idea that the gods intended that human beings – the king being the exemplary human being – should know the future. Regularity and repetition were built into the system and, in that way, events followed events in a certain 'readable' fashion, lending a logical predictive structure to omen divination. Even though we might say that omen divination did not predict but rather indicated possibilities, the practice was implemented for purposes of foresight, which is to say, it facilitated the Assyrian king's desire or need to project himself and his political agenda into the future with the best chance of success. The fact that magical means to avert or annul portents was possible only shows that the final arbiters of all things were the gods. And because the future was imagined as bound to the past through the repetition of signs and portents, action within a world so determined by divine design depended upon the ability of the king, through his diviners, to comprehend, observe, and interpret that design in the repeatable patterns, the language, so to say, of ominous phenomena.

Celestial and Terrestrial Omens

Divination from omens was not the invention of the Assyrian Empire, nor did it end with the Empire's fall. It was already an age-old practice, beginning in Babylonia during the second millennium BC. A poetic text from this early date is one of the earliest attestations to the use of stars as signs. It is a prayer to the 'gods of night' (*ilū mušītim*), the gods, that is, the celestial bodies in question, include some constellations we can identify, such as Canis Major, called Bow (*Qaštu*), Boötes, called Yoke *(Nīrum)*, Orion, called 'The True Shepherd of Anu' (*Šitadallu*/SIPA.ZI.AN.NA), Ursa Major, called 'The Wagon' (*Eriqqu*), and 'She-Goat' (*Enzu*) was the constellation Lyra. Already at the time of the prayer to the gods of night, in the period of Hammurabi of Babylon in the eighteenth century BC, celestial phenomena were compiled into lists of ominous signs, focusing in that early period on the moon, particularly on the lunar eclipse.

Unlike with the stars or the moon, where the celestial gods gave signs by their very appearances, extispicy, or the reading of the sheep's exta, required that an offering and a formal petition first be made to the patron gods of divination, Šamaš and Adad. As in the following prayer to be said before an inspection of the liver, the gods were conceived of as divine judges. The diviner prepared to go before the gods Šamaš and Adad, saying:

> Cleansed now, to the assembly of the gods draw I near for judgment (*ana dīnim*). O Šamaš, lord of judgment, O Adad, lord of prayers and acts of divinatory inspection (*bēl ikribi u bīri*), in the ritual I perform, in the extispicy I perform, place the truth! (*ina ikrib akarrabu ina têrti eppušu kittam šuknam*).

The events portended by omens from the liver as well as from the heavens applied to the king, the country, and the populace at large, covering all manner of military and economic disasters, the rising of floodwaters to destroy crops, plagues of locusts or of disease among the herds or the population as a whole. These phenomena were systematically arranged in lengthy lists, some that explicitly echo the diviner's guidebook in its reference to 'the city, king and his people.' An omen from the celestial omen list entitled *Enūma Anu Enlil* concerns a solar eclipse occurring on the 30th day of the month of *Ulūlu*, as follows:

> If on the 30th day (there is a solar eclipse): the city, king and his people will be well, (but) lions will become wild and block passage on the road. [EAE 33 § VI.13]

Or,

> If on the 18th day: The city, king and his people will have peace; (but) plague will destroy the herds. [EAE 33 § IV.8]

One tablet of *Enūma Anu Enlil* contains relatively detailed descriptions of lunar eclipse phenomena with portents for the various enemies of Babylonia, the point of view from which the omens were originally constructed:

> If an eclipse occurs on the 14th day of *Tebetu*, and the god (that is, the moon), in his eclipse, becomes dark on the east upper part of the disk and clears on the west lower part; the west wind (rises and the eclipse) begins in the last watch and does not end (with the watch); his cusps are the same (size), neither one nor the other is wider or narrower. (You should) observe his eclipse, i.e., of the moon in whose eclipse the cusps were the same, neither one being wider or narrower, and bear in mind the west wind. The verdict applies to Subartu. Subartu and Gutium . . . brother will smite brother; the people will suffer defeat(?); there will be many widows; the king of Subartu will make peace with the lands . . . Thus is its sign and its verdict.

This passage instructs the diviner just how to interpret the lunar eclipse in question, as pertaining to the region of Subartu in northern Mesopotamia, the region that would ultimately become Assyria. What is referred to as the 'verdict' refers to the social consequences of this eclipse as though rendered by the gods as judges in a case, evoking the same conception of the gods as judges found in the prayer to be said before reading the omens in the liver.

Celestial omens from *Enūma Anu Enlil* appear frequently in the correspondence between diviners and the kings Esarhaddon and Assurbanipal. As in the following passage from one such letter from an unknown scholar the correspondence affords us a glimpse of how the experts interpreted celestial signs and explained them to the king. The writer first quotes the omen: 'If Mars appears in the month *Ajaru*, there will be hostilities. Affliction of the *Umman-manda*.' He then explains: '*Umman-manda* means the Cimmerians (referring to a people of unclear origins but found in Anatolia south of the Black Sea).' He continues: 'The solar eclipse which occurred in the month *Nisannu* did not afflict the region of the Northland. Also the planet Jupiter retained its position. It was present for 15 days. That is propitious.' And again he quotes an omen: 'If the sun rises amidst a resplendent cloudbank, the king will become angry and raise his weapons,' and explains, 'As regards the rains which were (so) scanty this year that no harvest was reaped, this is a good omen pertaining to the life and vigor of the king, my lord.'

The parallelism between Mesopotamian liver and astral divination underscores the fact the signs in heaven were not causes of future happenings, nor did they affect the future by stellar influence, as we generally assume to be the case in astrology. Instead, looking to the future in ancient Assyria depended upon knowledge of a pattern of events in the past. Regular recurrence was part of the gods' design, or scheme, of things. That the Assyrians saw the future in perpetual negotiation with the past is demonstrable precisely by the evidence for divination and is a critical element in the Assyrian view of the world as a whole. All things held together as part of a divine design, whether temporally in the connection between past and future or spatially in the unity of the imperial lands and, at least in principle, the coextension of the Empire with the inhabited world. This divine design, in Akkadian literally 'drawings,' or 'plans' (*uṣurāti/* GIŠ.HUR.MEŠ), was instantiated in the ominous signs.

In the Reign of Esarhaddon

More detail about the implementation of institutional foresight in the Neo-Assyrian period can be found for the reign of Esarhaddon (681–669 BC). According to the Assyrian imperial master narrative, Esarhaddon told of his own appointment as crown-prince by divine selection. He related how his father, Sennacherib, 'questioned the gods Šamaš and Adad by divination and they answered him with a firm "yes", saying "He is your replacement".'

Esarhaddon's succession was troubled as he faced the subterfuge of his own brothers and would-be usurpers. He tells of his prayers to the gods Assur, Sin, Šamaš, Bēl, Nabû, Nergal, Ištar of Nineveh, and Ištar of Arbela, who, as he says, 'accepted my words ... with their firm "yes",' they were sending me reliable omens, (saying): 'Go! Do not hold back! We will go and kill your enemies.' And when in fact he claimed his throne and entered Nineveh, he said 'the south wind, the breeze of the god Ea, the wind whose blowing is favorable for exercising kingship, blew upon me. Favorable signs came in good time to me in heaven and on earth.' These inscriptions establish Esarhaddon's succession as the result of divine

166

selection, confirmed by divine favor, indicated by signs in heaven and on earth. Indeed, the announcement of the omens was an essential ingredient of the master narrative, the key to securing a legitimate claim.

One of the first of Esarhaddon's acts as king was his entry into Babylon to establish his kingship there. No Assyrian overlordship of Babylonia could be successful without formal acceptance and approval of the king by the Babylonian national god Marduk. In order to achieve this, Esarhaddon would have to restore Marduk's temple, the Esagil, which had been destroyed when Sennacherib, Esarhaddon's father, mounted a vicious campaign against Babylonia and their Elamite allies. After smashing the divine statues in the temples of Babylon and seizing their property, Sennacherib described his final destruction of the city:

> The city and its temples, from its foundation to its walls, I destroyed,
> I devastated, I burned with fire. The (inner) wall and outer wall, temples
> and gods, the ziqqurat of brick and earth, as many as there were, I razed
> and dumped them into the Arahtu canal. Through the midst of that city
> I dug canals, I flooded its ground with water, and the very foundations
> thereof I destroyed. I made its destruction more complete than that by
> a flood. That in days to come, the site of that city, and its temples and gods,
> might not be remembered, I completely blotted it out with (floods of)
> water and made it like a meadow.

When in 681 Esarhaddon became king and set out to establish his own sovereignty over the beleaguered Babylonia, he declared that the astral manifestation of the Babylonian god Marduk sent him a positive omen:

> At the beginning of my kingship, in my first year ... good signs were
> established for me in heaven and on earth concerning the refurbishing
> of the gods and the rebuilding of shrines. Jupiter (the astral
> manifestation of Marduk) shone brightly and came near in Month III
> and stood in his place where the sun shines. He reached the place of
> secret (where he was at his most propitious) for a second time in
> the month 'Opening the Door' (Month VII) and stayed in his place
> (i.e., Jupiter was stationary).

Without the celestial omen from the Babylonian god Marduk, Esarhaddon's reopening and reentry into the temple Esagil as King of Babylonia would have been impossible to sell. He was, however, not deterred, because, as the annals continue: 'In order to triumph (and) to show overpowering strength, he (the god Marduk) revealed to me good omen(s) concerning the re-entering of Esagil. The stars of heaven stood in their positions and took the correct (literally the "true") path (and) left the incorrect (literally, the "untrue") path.'

Another of the annals reiterates and expands this astral communication of divine approval, thus:

> In [order] to give the land and the people verdicts of truth and justice, the gods [Sin and] Šamaš, the twin gods, took the road of truth and justice monthly. They made (their simultaneous) appearance regularly on days [. . .] and (day) fourteen. Venus, the brightest of the stars, was seen in the west, [in the Path] of the Ea-stars. Concerning the securing of the land (and) the reconciliation of its gods, it (Venus) reached (its) hypsoma and then disappeared. Mars, the giver of decisions on the land Amurru, shone brightly in the Path of the Ea-stars (and) it revealed its sign concerning the strengthening of the ruler and his land.

For Esarhaddon to have advanced into Babylonia without knowing the favorable assent of the gods delivered through ominous signs in heaven would have been an arrogant sacrilege. Indeed, the *topos* of the king fallen from grace for ignoring omens, was known in Esarhaddon's time. It was perhaps best exemplified in the figure of the Akkadian ruler Narām-Sin from the third millennium BC, whose legend was part of the holdings of the Assyrian royal tablet collection at Nineveh.

Similarly for other acts of political gain, records of the appeal to the sungod Šamaš show that nothing of any real consequence would be undertaken without ascertaining knowledge of its outcome by divination. In the text of one such query to the sungod, Esarhaddon's plans for a campaign to the north of Assyria are set before the god, to be followed by an extispicy. The query reads:

> [Šamaš, great lord, give me a firm positive answer to what I am asking you! From this day, the 22nd day of this month, *Simanu* (III), to the 21st day of the following month, *Du'uzu* (IV), of this year, for 30 days and nights], the stip[ulated term for the performance of (this) extispicy–within this stipulated term], will the troops of the S[cyth]ia[ns, which have been staying in the district of Mannea and are (now) moving out from the territory] of Mannea, (will they) strive and plan? Will they move out and go through the passes [of Hubuškia] to the city Harrania (and) the city Anisus? Will they take much plunder and booty from the territory of [Assyria]? Does your great divinity [know it]?

In addition to the divinatory queries, a sizeable archive of letters from members of the royal advisory staff, principally the celestial diviners and the incantation priests, is preserved from the reigns of Esarhaddon and his son and successor Assurbanipal. The celestial diviners were consulted on similar situations as were the specialists in reading the liver. Here is a letter from a celestial diviner, one Bel-ušēzib, to King Esarhaddon concerning the invasion of the Mannean territory, not far from the ambit of the Cimmerians, cited in the aforementioned extispicy query:

> To the king of the lands, my lord, your servant Bēl-ušēzib. May Bēl, Nabû, and Šamaš bless the king, my lord! If a star flashes like a torch from the east and disappears in the east: the main army of the enemy will fall . . . If the king has written to his army (saying): 'Invade Mannea,' the whole army should not invade; (only) the cavalry and the professional troops should invade.

In a subsequent letter Bel-ušēzib reported on the fall of Mannea to Esarhaddon's army. He said:

> I saw the crescent of the moon but the sun was rising . . . Whether it was a crescent, or 'whether it appeared on the 15th, or whether it will appear on the 16th, it is an evil portent, and it concerns the Manneans. Wherever an enemy attacks a country, the country will carry this evil portent. Now the army of the king, my lord, having attacked the Manneans, has captured forts, plundered towns and pillaged the open country.

Indeed, in a later version of the annals, composed three years before the end of Esarhaddon's reign, he called himself 'the one who scattered the Mannean people.'

The most ambitious long-distance military operation conducted in the expansion of Esarhaddon's Empire was his invasion of Egypt and subjugation of the 25th Dynasty under the Kushite King Taharqa (Figure 8.3), also known as Tirhakah from the Bible. In a well-known episode in the 2nd Book of Kings, Tirhakah is the ally of Hezekiah, King of Judah. Sennacherib mocked Hezekiah for 'relying on Egypt, that broken reed of a staff, which will pierce the hand of any man who leans on it. Such is Pharaoh King of Egypt to all who rely on him.' In this instance, according to the biblical writers, and Lord Byron, things ended badly for Assyria. The Assyrian may have come down like a wolf on the fold, but the angel of the god Yahweh

FIGURE 8.3   Sphinx of Taharqa

slaughtered 185,000 Assyrians that night, and left the widows of Ashur loud in their wail. Sennacherib went home to Nineveh, where he was killed by two of his sons, and Esarhaddon succeeded to the throne. On the other hand, Taharqa remained Pharaoh long enough to later become once again the target of Assyria's expansion into Egypt under Esarhaddon. The relevant extispicy query reads:

> Šamaš, great lord, give me a firm positive answer to what I am asking you! Should Esarhaddon king of Assyria, strive and plan? Should he take the road with his army and camp, and go to the district of Egypt, as he wishes? Should he wage war against Taharqa, king of Kush and his army?
> In waging this war, will the weapons of Esarhaddon, king of Assyria, and his army, prevail over the weapons of Taharqa, king of Kush, and the troops which he has? ... Will Esarhaddon, king of Assyria, return alive and set foot on Assyrian soil? Does your great divinity know it? ...

> Be present in this ram, place in it a firm positive answer, favorable designs, favorable propitious omens by the oracular command of your great divinity, and may I see them. May (this query) go to your great divinity, O Šamaš, and may an oracle be given as an answer.

We can only imagine that Šamaš did write a favorable answer in that liver because Esarhaddon's victory stele relates that

> after the god Aššur and the great gods, my lords, had ordered me to march far along remote roads, (through) rugged mountains (and) great sand dunes, where (one is always) thirsty, I marched safely (and) in good spirits. As for Taharqa, the king of Egypt and Kush, the accursed of their great divinity, from the city Ishupri to Memphis ('Mempi'), (his) royal city, a march of fifteen days overland, I inflicted serious defeats on him daily, without ceasing. Moreover, he himself, by means of arrows, I inflicted him five times with wounds from which there is no recovery; and (as for) the city of Memphis, his royal city, within half a day (and) by means of mines, breaches, (and) ladders, I besieged (it), conquered (it), demolished (it), destroyed (it), (and) burned (it) with fire.

This inscription, found at Zinjirli (ancient Sam'al) in southern Turkey (Figure 8.4) was erected as a victory stele in 671 BC to commemorate the destruction of Memphis. The small figure, Ushankhuru, the young son and crown prince of Taharqa, is shown in bondage. The opening statement of the stele is testimony to the gods' power to decree the future by sending signs and making judgements:

> The god Assur, father of the gods, who loves my priestly service; the god Anu, the powerful, the foremost, the one who called my name; the god Enlil, lofty lord, the one who confirmed my reign; the god Ea, wise one, knowing one, who decrees my destiny; the god Sin, shining Nannar, the one who makes signs favorable for me; the god Šamaš, judge of heaven and netherworld, the one who provides decisions for me; the god Adad, terrifying lord, the one who makes my troops prosper; the god Marduk, hero of the Igigu and Anunnaku gods, the one who makes my kingship great; the goddess Ištar, lady of war and battle, who goes at my side; the Sebitti, valiant gods, the ones who overthrow my enemies; the great gods, all of the them, who decree destiny (and) give victorious might to the king, their favorite.

Reference to the gods in the opening lines of the stele was surely to restate before them the piety of the king. But the pietistic account of political and military triumph is consistent with the other texts quoted

FIGURE 8.4 Zincirli Stele of Esarhaddon

before that reflect pragmatic behind-the-scenes operations, such as the extispicy queries to the sungod and the letters from scholar-scribes concerning celestial omens. These texts confirm for us that hymn-like rhapsodizing about the gods in annalistic texts was not merely a royal rhetorical flourish, but part of the structure of the master narrative, in which the gods were in charge of the future and could be called upon to let their decisions be known. Knowing what the gods had in store was then the basis for action and for ensuring the security and growth of the state and of royal power.

Conclusion

Those in power have perhaps the most to gain through a judicious use of foresight. It is therefore not surprising that the written remains of one of the most powerful, indeed the first, of ancient empires trained and employed teams of experts in the art and science of looking forward. The *literati* of ancient Assyria, the scribes and scholars of royal courts and chancelleries, developed an elaborate practice of foresight, analyzing, synthesizing, and interpreting the phenomena around them in accordance with a system of 'reading' the divine language of signs. The practice was both a tool for royal legitimation as well as to enable the king to make decisions as to how to project his goals into the future.

Ancient Assyrian kings were invested in the art of anticipating the future in the realms of politics and economics, much as we ourselves are invested in projecting and preparing for the future in our modern realms of business, government, and science. They conducted business with an eye on the omens, as we see from privileged glimpses within palace chambers through the correspondence between these kings and their advisors. In the tablets produced by the scholars ranging from handbooks containing lists of ominous signs, celestial and terrestrial, to divinatory prayers, queries, and rituals addressed to the patron gods of divination, in all these numerous and diverse texts, we gain insight into the context, practice, and purpose of Assyro-Babylonian divination and its role in the earliest and most fully documented cultivation of foresight. They tell of the investment in the future survival, continuity, and security of the state to its fullest extent as predicated on correct behavior toward the gods and in diligent reading of the gods' written messages concerning the future.

Essential to correct behavior toward the gods was the building and maintenance of temples. Indeed, the annals consist both of reports on military expeditions and on the building or restoration of temples. Each of these activities was equally geared to future security. In no uncertain terms Esarhaddon stated this when he said in reference to his restoration of the Ešarra Temple, the temple of the god Assur in the religious capital of Assur itself:

> I built (it) for my life, the prolongation of my days, the securing of my reign, the well-being of my seed, the safeguarding of the throne [of] my priestly office, the overthrowing of my enemies, the prospering of the harvest of Assyria, (and) the well-being of As[syria].

And in the closing of his account of the restoration of another temple, that of the goddess 'Queen-of-Nippur,' where he warned any and all who would destroy his work or his words, the focus on the future could not be more clear:

> If at any time in the future, during the days of the reign of some future ruler, this temple falls into disrepair and becomes dilapidated, may (that ruler) seek out its (original) emplacement (and) repair its dilapidated parts! The gods will (then) hear his prayers. He will lengthen (his) days (and) enlarge (his) family. (But as for) the one who by some crafty device destroys an inscription written in my name or changes its position, may the goddess Queen-of-Nippur, great lady, glare at him angrily and make his name (and) his descendant(s) disappear from every land!

Because we need to read the Assyrian royal annals for details of history and to reconstruct the empire's rise and fall, we focus less on the internal imperial narrative of the corpus, which concerned itself more with the divine selection and legitimation of Assyrian kings, as well as with proving in the sight of the gods the king's maintenance of the proper relationship between divine and king, thus ensuring the future continuation of the Assyrian state. The Assyrian imperial narrative found a way to tie past, present, and future together into a complex of ideas with real consequences in action. Providing a framework within which the Assyrian royal annals as well as the documentation of contemporaneous letters, omens, oracle queries, and extispicy reports all fit together, that narrative comprehended a future for which specific divinatory techniques were the mainstay of institutional foresight. Neither livers nor celestial portents would have had any contemporary significance apart from the framework provided by the Assyrian imperial narrative, geared as it was to the future and to the survival and success of Assyria. And this is one of the great puzzles of the cuneiform scholastic tradition, namely, that we do not know what the purpose of liver or celestial divination was after the fall of Assyria, for by then we have no more access to how those kinds of divinatory practice related to a world of political or social reality. Without the continuation of royal correspondence, annals of the deeds of kings, or the queries and reports to the gods, the cuneiform omen texts, which were preserved for many centuries into the Neo-Babylonian, Persian, and Hellenistic periods of Mesopotamian history up to the beginning of the Common Era, take on the character of a frozen remnant of foresight planning without an anchor in the real world.

Further Reading

Loveridge, Denis (2009) *Foresight: The Art and Science of Anticipating the Future.* New York and London: Routledge

(2001) 'Foresight – Seven Paradoxes,' *International Journal of Technology Management* 21, 781.

Oppenheim, A. L. (1974) 'A Babylonian Diviner's Manual,' *JNES* 33, 205

Marcus, Alfred (2009) *Strategic Foresight: A New Look at Scenarios.* New York: Palgrave and MacMillan

Goetze, A. (1968) 'An Old Babylonian Prayer of the Divination Priest,' *Journal of Cuneiform Studies* 22, 25

Index

Index

Printed in the United States
By Bookmasters